W9-AOC-716

The Paranormal

The Paranormal

A Scientific Exploration of the Supernatural

Arthur Ellison

Dodd, Mead. New York

To Marian, Jennifer and Richard:
a most helpful and supportive family.

First published in United States of America 1988
by Dodd, Mead
71 Fifth Avenue New York NY 10003

Published simultaneously
by Harrap Limited London

This book was conceived, edited and
designed by Thames Head a division of BLA Publishing Limited
TR House
1 Christopher Row
East Grinstead
Sussex RH19 3BT
United Kingdom

Editorial Director
Martin Marix Evans

Design and Production
David Playne

Editor
Gill Davies

Illustrators
Paul Beebee Craig Warwick

Designers and Assistants
Tony de Saulles Marc Langley Gail Langley
Miles Playne

Typeset in ITC Clearface on Scantext by Playne Books Avening Gloucestershire
processed by Townsend Typesetter Limited Worcester
Printed and bound in Great Britain by
The Bath Press

Library of Congress Cataloging in Publication Data
Ellison, Arthur J.
The Paranormal.
Bibliograph P.
Includes Index
1. Psychical Research. 1. Title.
BF 1031.E52 1988 133 88-399
ISBN 0-396-08893-7

CONTENTS

Acknowledgments

I am deeply grateful to The Society for Psychical Research — the first and only scientific body in the UK looking dispassionately at 'those faculties of man, real or supposed, which appear to be inexplicable'. Many friendly members of that Society (which has no corporate beliefs) have discussed with me matters considered in this book. I am grateful to them all (too numerous to mention by name) and for their books and papers, some of which I have listed. To the secretary, Eleanor O'Keeffe, I express special thanks for her never failing helpfulness. I am also grateful to The Theosophical Society and its equally friendly members (also having no corporate opinions except for the fact of human brotherhood) for stimulating reading and discussions over many years and for encouragement to examine and investigate.

Any errors in paraphrasing the work of friends of either society are mine and I apologise for them. I regret also any error of detail: I hope there are few, and none of substance.

Especially I owe a debt to the many gifted psychics and others who have helped me with research over the years. I have almost invariably found psychics to be transparently honest, helpful and friendly, ever ready to assist a serious investigator.

My warmest thanks go to Martin Marix Evans, whose initial discussion led to the book, to its editor Gill Davies, to Tony De Saulles, Marc Langley and the illustrator, Paul Beebee who greatly improved my efforts. Finally my gratitude is due to my wife Marian for encouraging support, and for careful typing, to my son Richard for his word processor expertise and to my daughter Jennifer for her interest.

Preface

This is my psychical research autobiography. As a scientist, I have tried to assess, as honestly and clearly as possible, a lifetime of experience with the paranormal. I must stress that I have genuinely experienced all the occurrences I describe.

This is, I believe, a most unusual book. I have always been interested in psychical occurrences and have been exceedingly lucky in my searches for firsthand experience of most phenomena and have diligently kept full notes, written either at the time or very shortly afterwards. I have been to untold trouble to follow up likely phenomena, including arriving home by the milk train for a year or more every Monday morning after attending Sunday-evening séances! I have devoted hundreds of hours to studying — by hypnosis, meditation and psychic development — unusual aspects of the mind. I have devoted a perhaps similar time to discussing the experiences of mediums and mystics and have examined Indian and other philosophies. I have also lectured to various international conferences on Western science and mysticism/religion.

My approach to the subject has always been a scientific one. I am not particularly interested in claimed phenomena for which there is no valid evidence; I agree with C.G. Jung when he wrote that he was not so foolish as to suggest fraud for every phenomenon which he could not understand.

Psychic research is much more exacting than many other scientific subjects. It forces one to examine the very basis of one's views about consciousness and the universe. Most of the difficulties met in accepting even well-verified phenomena are due to the absence of a complete theory — a model or paradigm — within which they fit. Most 'ordinary' Western science fits the model presented by naïve realism — it is all out there in three-dimensional space, and science is the process of examining it (without unduly affecting it). Because most psychic phenomena do not fit this model they are rejected by many so-called scientists: such phenomena are impossible, therefore they cannot happen, therefore they do not happen. However, quantum physicists and parapsychologists are now leading scientists into new realms, and providing an escape from this arid, unscientific philosophy.

I have twice been President of The Society for Psychical Research: this is a scientific body, the first in the world to study this subject, and has no corporate opinions. The Society gained quite a reputation over many years for excessive scepticism, but after a hundred years or so really should know a little about the subject — at the very least whether certain phenomena actually take place or not, disregarding non-comprehension of such events. When I became President I decided, in view of the growing strength of evidence, to set a precedent by coming off the fence and by saying that some things I really do know with a 'normal' degree of certainty. (Nothing in this whole wide world is one hundred per cent certain.) I set out what I considered we really knew after a century of work and was astonished to find out how many people agreed but felt it unwise to put their views into published form.

The United Kingdom now has its first university chair in parapsychology. I am sure that all open-minded scientists and thinkers will have wished the new professor every success. I hope that sufficient extra research funds will become available to allow the many well-qualified young people who wish to research and take higher degrees in the subject such an opportunity.

In the meantime considerable progress is being made. The media no longer ask me for the names of sceptics and believers for a programme on this subject: they ask for well-qualified critical people with some experience of the paranormal. My own profession, electrical engineering, has recently been enjoying wide-ranging, fruitful and constructive discussions of the evidence. The best modern evidence involves the artifacts of electrical engineering; noise sources, microprocessors, visual display units and the like. The days of the amateur are, sadly, almost at an end in this regard. But the subject is so wide-ranging and spans so many other subjects, from engineering and physics to psychology and philosophy, that we are all amateurs in some regions. Multi-disciplinary teams are needed to study it properly.

The subject is so challenging and fundamentally important that it must be studied by the ablest and most intelligent among us. It is very clear — at least to me — that a major paradigm shift is on the way: the very way we look at the world around us and at our own minds, is due to change. The way we interpret the phenomena we experience — and what qualifies as fact — must now be reassessed. This is likely to have an impact upon our approach to medicine and religion, as well as to conventional science. The way does now seem to be open for serious consideration by scientific people of religion — in the widest sense — but that is too large a subject for this book to evaluate in depth.

The final chapter provides some new and stimulating ideas which may assist us to reach at least a rough concept of ourselves and the universe, if not a full understanding. I am sure that the information therein is not wholly consistent, not wholly logical and certainly open to many criticisms — but it will provide a starting point. We do not, indeed cannot, know all the answers. But we all surely must make some attempt to solve the major problems of life and existence. I suggest that parapsychological research is a good way for a modern scientifically-educated person — or anyone whose culture is grounded in science — to proceed from hard materialistic science to something deeper, subtler and much more meaningful.

I hope that this book will provide an enjoyable and stimulating means of considering the evidence for the paranormal as well as our thoughts and responses to it.

1
PRESENTING THE PARANORMAL

What would you think if someone you knew well to be trustworthy and normally accurate told you that he had last year been involved in a serious accident, that his heart had stopped beating, his breathing had ceased and a doctor had pronounced him to be clinically dead? After coming on the scene, it had taken the doctor a full five minutes to restart the heart and breathing and during this time your friend had undergone, not a period of blankness, but a most wonderful experience involving moving down a dark tunnel and coming out into the light.

There he had had a review of his life in pictures and had been helped to evaluate this by an 'angel-like' being. Next he had found himself in beautiful countryside and had seen his 'deceased' mother and father looking young and happy, apparently waiting to welcome him at the other side of a river. Then all at once he heard a voice say, 'It is not yet time,' and suddenly found himself conscious on the ground, wracked with pain, the doctor having finally succeeded in resuscitating him.

Again, suppose an old lady told you that during the war she was sitting at her knitting when she suddenly saw standing in the doorway, in full uniform, her son whom she thought was away serving in the army. Jumping up to welcome him she found that he had unaccountably disappeared and she was unable to locate him anywhere. Two weeks later she received a telegram to inform her that her son had been killed at the front at about the time she 'saw' him in the room with her.

A third example: a friend of yours has had a regular severe migraine every three weeks or so for some years. Regular visits to the hospital, consultations with more than one doctor and the taking of several different drugs have made little impression on the headaches. Then one day she is persuaded to go and see a 'spiritual healer' — much against her inclinations as she cannot see how it will be of any use. The healer, she explains, simply placed his hands against her head for some five minutes. This was three months ago and your friend has not had a migraine since, though nothing else has altered in her life.

Finally, an acquaintance tells you in conversation that a week ago he was sitting in his armchair reading and thinking of going to bed. Suddenly he had a strange feeling of 'moving out' and found himself floating across the room. To his great surprise he was apparently able to see himself still in the armchair and still reading. His first thought, he explains, was that he had suddenly died. However, in a few minutes he had another strange feeling, this time of being attracted towards the reading body, and found himself again back to normal with no unpleasant after-effects. He tells you that he was quite certain that he had not fallen asleep and that there was no break in his consciousness from beginning to end.

These are four examples of the kind of occurrences dealt with in the pages to come. What do you make of this kind of story? Do you dismiss them all immediately as 'imagination'? Or do you think that as the people who have told you these things are

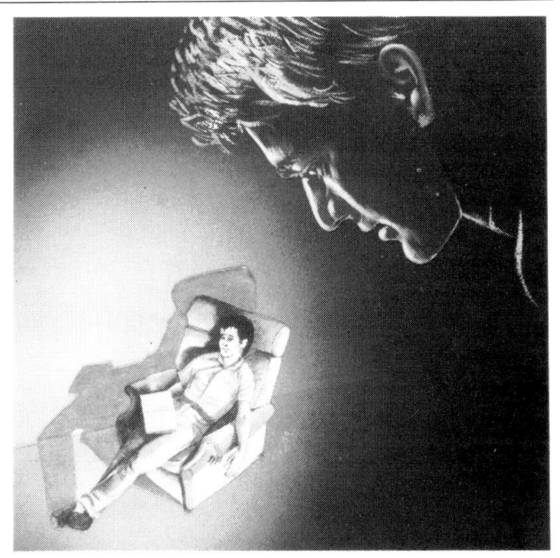

normally reliable that these events cannot be just hearsay, old wives' tales, they must be considered seriously? And if so, exactly how do you go about considering them?

These four examples are typical of many thousands of similar cases. In the chapters that follow we shall consider further even stranger and more astonishing accounts but, of greater importance, these accounts will be subjected to a careful and scientific investigation. We shall be considering how the methods of science — the rigorous, objective, unbiased methods that have led to so much material benefit of every kind — can be applied to such phenomena. Certainly experiences of that kind do not appear to fit the scientific picture of the universe that we were all (at least those of us brought up in the West) taught at school. Must they therefore be rejected — as religion is so often rejected — as being the delusions of an unbalanced or temporarily deranged mind? Should we accept the rejections of those scientists whose response is: 'Stuff and nonsense! You don't believe all that rubbish do you? Such occurrences are impossible; the person telling you that has been deluded by himself, or by others. Have some common sense!'?

What is wrong with this 'scientific opinion'? The answer is that it is grossly unscientific and shows the so-called scientist to be lacking in the most elementary virtues and understanding of his or her profession. The last thing a true scientist could possibly be is prejudiced concerning evidence which he or she has not properly examined. But of course that does not mean that the normal 'rational' explanations of claimed occurrences should not be looked for first. The mind should be open, not empty. This then is what we shall attempt to do. But where the 'normal' explanation does not suffice we must not be so unscientific as to reject the facts. The facts come first; the theory comes second. If our present collection of theories does not fit all the facts of human experience — when we are of course certain that they are genuine human experiences — then so much the worse for the theories. Such theories must be adapted, extended, rejected, recast until they do fit. That is what science is all about. It is not good enough simply to reject repeated evidence from reputable people as 'imagination', or delusion. It is also not good enough (and this, unfortunately, is the frequent reaction of such 'scientists' to religious problems) to reject the deepest and most significant questions that arise in the human mind, as being devoid of meaning.

WHAT IS REALITY?

We shall find as we come to consider examples of the paranormal that we are up against problems of reality and perception. We shall find ourselves faced with such fundamental questions as 'What do we mean by reality?' Are the five known senses the only ways of acquiring information? What about the tacit assumption of many, perhaps the majority of Western scientists, that a human being is no more than a few kilograms of tissues with a little computer at the top; is this really valid? Similarly, what of the view that there is little more to the universe than objects distributed around us in space moving through time; is this adequate?

We shall see, as we proceed, that the kinds of experience described, in addition to the views of modern particle physicists, are showing cracks opening in this simplistic picture of the universe. We shall begin to see ways in which (though this is not the subject of this book) religion might now be seriously considered as an integral part of the pattern. We shall see how the rigours of science are showing that the basic assumptions of western science are quite inadequate. We shall see that human beings are vastly more than the materialist hypothesis suggests: that there are other, and important ways of acquiring information than through the five senses — that our ordinary life is but one aspect of a very much greater whole. That 'naïve realism'(or seventeenth-century Cartesian dualism which later became positivism), must now be rejected.

THE BASIS OF SCIENCE

Many people think that science is the business of describing the physical world 'out there' with ever increasing accuracy. That scientists are, with the aid of various tools such as microscopes and telescopes, together with all the paraphernalia of the modern scientific laboratory, achieving ever greater accuracy in their pictures of the physical world. This view of science is completely, utterly and fundamentally wrong! In fact science is instead merely the process by which we build mental models to represent our experiences. It is important to pursue this a little further because a clear understanding will be essential as a basis for comprehending much of the material to come later.

Newton observed the movements of the heavenly bodies and also, so it is said, the falling of an apple to the earth. He devised a mental model representing all this which involved objects attracting each other in a way which depended on their masses and the distance between them. He wrote his so-called law of gravitation. That law was used without problems for some three hundred years. However, at about the beginning of the present century when more accurate measurements were made of the orbit of the planet Mercury, predictions using Newton's law were found to be inaccurate: eventually Einstein came along and devised a quite different gravitational law — another mental model

— representing the old plus the new facts. In Einstein's hypothesis all masses distorted the space-time continuum and objects floated freely through it. There was no idea of their attracting each other. It is perfectly clear that these so-called laws of nature are actually only mental pictures. The mental pictures represent human experiences. Newton's law is still used most of the time and gives a good approximation but when great accuracy is needed Einstein's law is used. Thus when the mental models cease to be useful in the prediction of future experiences they are discarded for new models. There is nothing permanent about these 'laws' of nature.

It is important to realize that all we have for certain is experiences in our minds (however we may define the mind). All of us when we were babies gradually made sense of these mental experiences by ordering them into the physical world. The way we look upon the physical world (that is the mental model we have of those experiences to which we can attach words — tables, chairs, other people, mountains or whatever) depends upon our culture, on our parents, our education and so on. People of quite different cultures, such as bush-men or aborigines, look upon the physical world in quite different ways from those of us brought up in a Western science-based culture.

The American philosopher F.S.C. Northrop has useful things to say concerning this. He refers to two kinds of knowledge: *aesthetic knowledge* means the experiences we have in our minds before we have 'made sense' of them and *theoretic postulated knowledge* refers to the objects of the physical world including our own bodies and senses. It is important to realize that we have 'postulated' these objects all around us to make sense of the mental content. The various comparisons we make between the input of different senses Northrop refers to as 'systemic correlation'. The whole picture hangs together very well and normally we have been so conditioned to all this that we consider the physical world to be 'out there' quite independent of ourselves, and that science is the process of examining and describing it, without of course altering it. As we have seen and shall see further, these latter ideas are fundamentally wrong and will be shown to be so through the experiences and mental models of particle physicists (in the microstructure of the universe) and the experiences and mental models of psychical researchers or parapsychologists (in the macrostructure).

A MODEL OF THE HUMAN MIND: AN ICEBERG CONCEPT

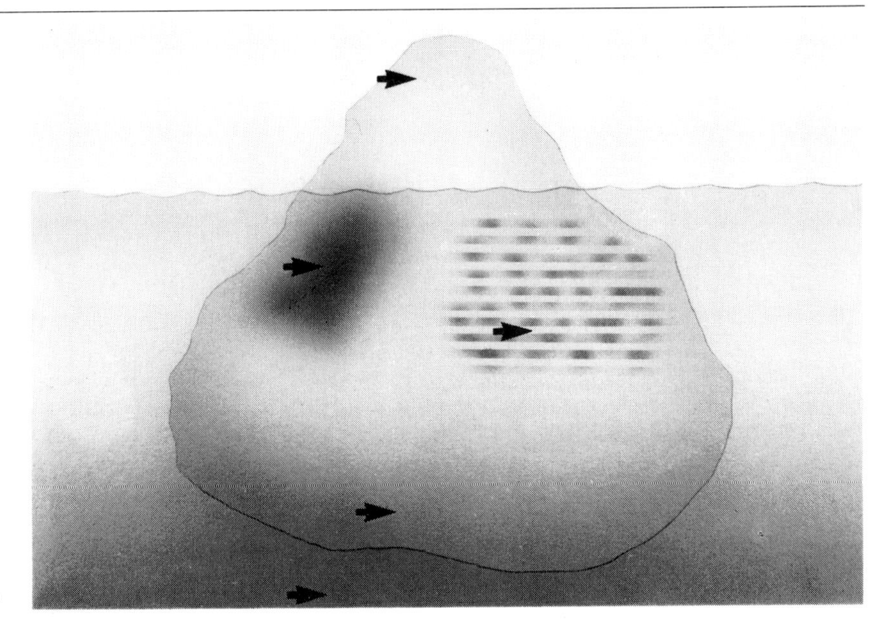

The diagram on page 11 shows a crude model of a human mind which, by and large, appears to fit most of the experiences called psychic. The diagram shows an iceberg floating in the sea. This represents a human mind. The part of the iceberg above the sea represents the conscious mind (the part you are using at the moment as you read and think) and the part below the sea represents the personal unconscious or 'subconscious' mind. The iceberg below the sea also contains the memory store which is shown as though it is rather like a computer store. There is within the unconscious mind machinery for selecting the memories we require as and when they are needed and pushing them up into the conscious mind. (The unconscious mind is most certainly not 'unconscious' as it is functioning both awake and asleep — so it is a pity we appear to be stuck with the term unconscious.)

This machinery of the mind is much more like an animal than a machine: I call mine George. George is that part of the mind which, from our memories of experiences (and maybe from other material too) dramatizes our dreams when the conscious mind is quiescent in sleep. George can be trained to wake you up at a specific time in the morning (he seems to have some sort of biological clock). George is a-logical (that is he has no conception of logic as does the conscious mind). Sometimes he associates two quite disparate experiences and permanently links them together. For example, he might link walking on to a stage with a freezing up of the vocal chords. The result will be that a trained singer can find it impossible to sing in front of an audience even though there is no problem in singing beautifully at home or with the teacher.

Your psychiatrist who is trying to discover why you have certain unpleasant and inconvenient bodily symptoms, even though there is nothing physiologically wrong, will be very interested in the dreams which George produces; this will provide clues to the wrong associations which George has somehow acquired. In hypnosis, the hypnotist places the conscious mind in an uncritical relaxed state and talks directly to George. In this way all sorts of phenomena which appear to be 'out there' can be produced, such as extra people who are not physically present. The hypnotist can also 'remove' people who are in fact physically present by appropriate suggestions to George (assuming the subject to be a good deep trance subject). Such a subject will have blisters produced by George when the hypnotist states that a pencil is a red-hot poker and touches the arm of the subject with it.

This aspect of the iceberg model will be found useful when we come to consider telepathy, healing and ostensible evidence for human survival beyond the death of the physical body.

As the consciousness is imagined to focus further and further down the model, time and space must be imagined to become quite different from those concepts with which we are familiar in our normal waking consciousness. At the deepest levels of the model (that is in the depths of the sea which corresponds to the 'collective unconscious') the consciousness is as it were in the 'eternal now' — past, present and future being all rolled into one, as is every aspect and part of the so-called physical world. This is the area of mystical experience.

Finally, concerning this model of the human mind, we can imagine that if a person is psychic there is somehow a crack in the iceberg which allows material normally in the unconscious mind (below the sea line) to float up into the conscious mind. Thus psychics have a rather wider range of experiences than the rest of us and they postulate various mental models to account for these to include such things as the astral plane, an astral body and so on, all terms to describe the mental models used.

Information from the unconscious is able to flow up through the 'crack' in the psychic's mind.

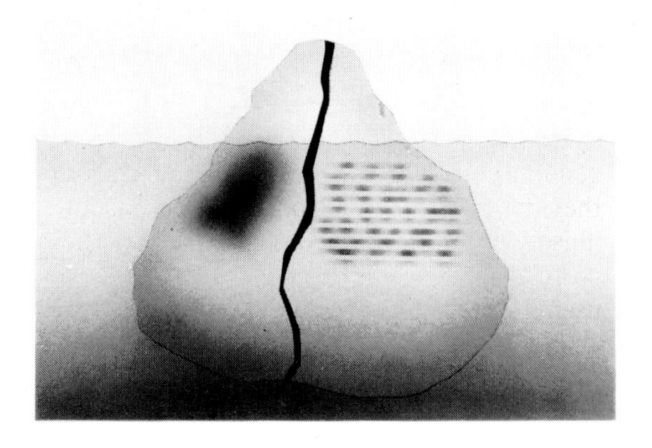

The human mind, which dominates our behaviour, responds to circumstance. We are different people in different situations and the reality of 'who we are' is very complex.

That is all we need say for the moment about our iceberg model of the human mind: however, it will often be referred to in later pages.

The rather oversimplified idea of the physical world as being distributed in space all around us and being quite independent of ourselves will be called 'naïve realism'.

We began this introductory chapter with a description of four cases: these were just typical examples of many which have been recounted as genuine human experiences — and which indeed they truly are. We then considered the somewhat illusory nature of the so-called physical world and postulated a very crude mental model which will be useful in the organizing and understanding of psychic and other experiences. Later I shall give a number of illustrations to show that there is not a great deal wrong with that representation of the mind: it explains much more than do some other perhaps even vaguer representations. It is important to realize that the mind is not to be considered as in some way in physical space. Physical space is just one example of the 'theoretic postulated knowledge' Northrop described. (See page 11)

You will find that it is quite difficult to think in terms other than those of naïve realism. We are all naïve realists most of the time and this model of the universe works well for most of our experiences. We have been conditioned by many years of thought and experience and are thoroughly habituated to naïve realism. This will not be the right approach to the subject raised by this book. Most of the questions parapsychologists are asked are based on naïve realism. It is therefore important to be aware of its influence and to try and formulate a more suitable model for what is happening.

2
APPARITIONS

Probably 'ghosts' (or 'apparitions') is the first thought which occurs to many people when they come across the words psychical or parapsychological. The commonest question psychical researchers are asked by the media (and others) is, 'Do you believe in ghosts?' A former president of The Society for Psychical Research, G.N.M. Tyrrell, pointed out many years ago that this was just about the most ambiguous question one could be asked. He suggested that if it were reworded in the form, 'Do you believe that people sometimes experience apparitions?' then there would be no problem. The answer then will be, 'Yes, of course they do'. Two questionnaires circulated in the early days of the SPR showed that about one person in ten has an experience once or more during their lives of perceiving an apparition. There is no reasonable doubt about this. But what the questioner usually has in mind when asking 'Do you believe in ghosts?' is 'Do you believe that dead people walk about the earth sometimes in their next-world bodies?' The answer to that question must be, at least in my case, I very much doubt it.

The 'haunting' type apparition is actually only one of five principle types of apparition. In this chapter we shall consider those five principle types, looking at examples of each and exploring how the model of the human mind which we sketched in the first chapter may be applied in order to 'understand' the phenomena.

FIVE PRINCIPLE TYPES OF APPARITION

Haunting type

The type of apparition which is supposed habitually to haunt certain places has been considered from earliest times to be a subtle part of a human being, persisting after death and returning to earth for some particular purpose, for example, to correct some wrong done. This does not seem in any way to accord with the facts. To the percipients the haunting-type apparition is much more like a piece of photographic film — of a shadowy person being run through some mental projector. The figure seems to be 'out there' in physical space but usually appears to do exactly the same things every time it is perceived; it seems to go through the same actions, trace the same path in space, disappear in the same way, and so on. It is fairly rare for it to appear to have any consciousness of being observed. It is more like a residuum of a person carrying out some sort of automatic actions rather than a surviving dead person who has for some reason returned.

The sceptic will refer to such experiences as 'imagination', but they are far from this. We are all familiar with the imagination which we deliberately use to construct pictures in the mind. However, in the case of the perception of an apparition, it is as though something real is presented to us in ordinary physical space and we observe it. We have no control over it whatsoever. It is not possible to photograph the apparition nor, in those rare cases where words are spoken or sounds made, is it recordable. In other words it is, in the definition of a psychologist, an hallucination. Calling it an hallucination of course explains nothing.

It is not my intention to provide large numbers of cases: there are many in various excellent books. I will however give one or two of each type of apparition as a typical illustration and will apply myself mainly to discussing possible explanations. (There are a number of cases which do not fit easily into any of the classifications considered here but in general most do.)

The Morton Haunting at Cheltenham

This is one of the best evidenced cases on record. It occurred early on in the history of The Society for Psychical Research at Cheltenham where F.W.H. Myers lived (one of the founders of the SPR); a reporter, having the pseudonym in the records of Miss R.C. Morton, was encouraged and helped by him. An excellent description of the case is given by Andrew MacKenzie. The concerned family consisted of Captain and Mrs Despard, Rosina Despard (aged nineteen at the beginning of the case and being educated as a doctor — making her quite a pioneer in 1882), Edith Despard aged eighteen, Lilian fifteen, Mabel thirteen, Henry sixteen and Wilfred who was six.

The figure of a weeping woman dressed in black was seen by many people over a period of eighty years.

The apparition was of a woman dressed in black; she was tall and held up a handkerchief to obscure her face. She was observed by a score of people over a span of about eighty years. The first experiences occurred in 1882 and Myers interviewed the family and staff some two years later. He found a remarkable uniformity in the experiences of the witnesses, even in cases where there had been no communication between them.

The house was detached and in 1882 modern (about twenty years old) and in a good state of repair. The figure first seen by Rosina was as described and standing at the top of the stairs. It soon descended and was followed by Rosina whose candle then burned out. The impression was of a lady in widow's weeds. In two years Rosina saw the figure about six times with perhaps half that number of sightings by others. Her sister first saw it go into the drawing room and thought it to be a visiting nun. She enquired of the family and was told that there was no one like that in the house. A housemaid later saw it and her description agreed with that of the others. Later Rosina's younger brother and another boy observed it from the garden in the drawing room and ran in to see who was crying. In later sightings Rosina followed the figure into the drawing room where it stood by the window, later moving along the passage to the garden door, where it disappeared. On one occasion Rosina stood quietly by it in the drawing room and asked if she could help. The woman looked as though she might speak but did not, and moved to the door. Rosina made a further attempt to speak to her but gained the impression that she was unable to speak. Then the figure disappeared again by the garden door. Rosina occasionally heard faint noises of gentle pushes on her bedroom door and light footfalls.

On one occasion Captain Despard and Rosina's sisters were sitting in the drawing room when Rosina came in. Soon the figure entered and stood behind the couch, being observed very distinctly by Rosina but by no one else. (The younger brother was not present.) Again, after half an hour, the figure walked towards the garden door and disappeared, after stopping as though trying to speak. The cook and Rosina's three sisters all heard the footfalls on a later occasion when in their bedroom on the top floor and they were heard also by Rosina's married sister on the floor beneath it. Whenever Rosina heard the footfalls and looked out of her bedroom door she saw the apparition. The cook said that the footfalls were unlike those of anyone else in the house and she described the same figure, having seen her once during the night.

A neighbour opposite, General Annesley, had noticed a lady in his orchard crying who fitted the same description. He thought she was Rosina's married sister in mourning for her baby son. The figure looked quite normal and solid.

Many other people, including new and uninformed servants, had similar experiences. Six or seven years after the first report the figure was rarely seen though noises were heard. These also faded. Rosina did several tests, seeing the figure walk through cords she stretched across the stairs. She never succeeded in touching it as it always eluded her or disappeared: it often appeared in rooms with closed doors. Two family dogs appeared to see the figure and to be frightened by it.

Andrew MacKenzie obtained evidence of a figure like this apparition, complete with handkerchief held to the face, being seen in a building opposite the Despard's house some eighty years after the first sighting. The Cheltenham ghost is known to the psychical researchers of today as an excellent, competently researched and typical case of a haunting-type apparition, showing well the significant features.

The Johnnie Minney case

This is a more recent case of an apparition seen by an Australian lady Mrs Stella Herbert who, soon after arrival in England, went to stay at a farm in Huntingdonshire with another Australian lady Mrs Shirley Ross. The farm was also separately occupied by an English woman Miss Margaret Minney, who had lived there all her life and met Mrs Herbert only very briefly on arrival.

During the first night of her stay Mrs Herbert was awakened by a little boy kneeling beside her bed. She described him as having a pleading look, a thin and drawn face and fair straight hair. She had the impression that he was tall and bony. He did not speak but she felt that he wanted her to call his mother. When she called 'Mummy' loudly the boy disappeared.

The boy looked very ill and seemed to want Mrs Herbert to call his mother.

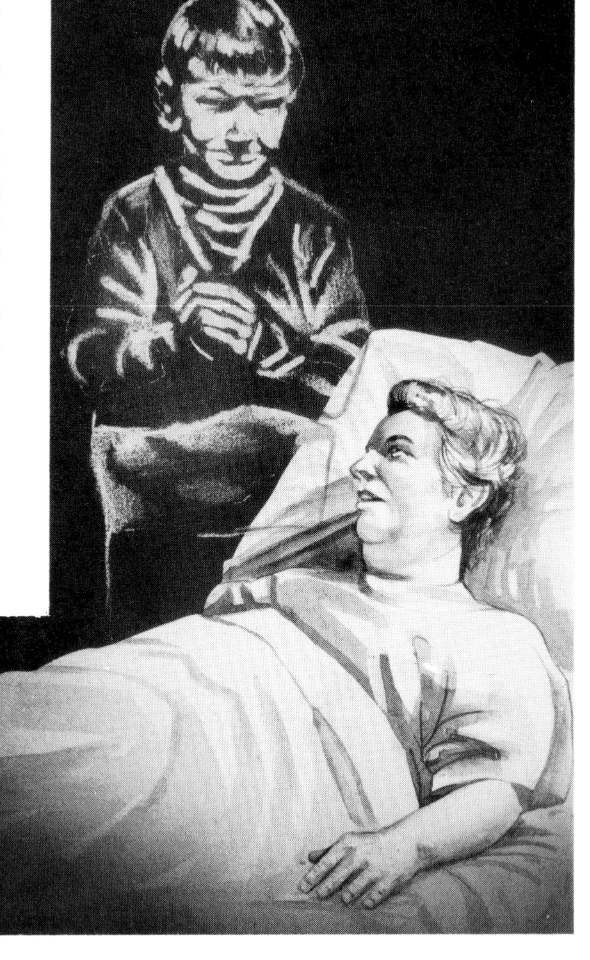

Soon afterwards Mrs Ross asked Miss Minney if a small boy had ever died in that house.

'Yes, my brother Johnnie,' she replied.

Mrs Ross said, 'Come to my room and listen to Mrs Herbert.'

Mrs Herbert recounted her experience to Miss Minney especially describing the emaciated appearance of the boy.

Miss Minney remained silent until the end and then said 'But I know that was Johnnie, my brother, who died when he was five.'

Miss Minney told the two ladies that her brother Johnnie had been sleeping in the room occupied by Mrs Herbert when he was taken terribly ill with meningitis in February 1921. She remembered his crying out with pain and often shouting 'Mummy'. His parents were away in London at that time. On their return Johnnie was moved to the room to be later occupied by Mrs Ross, where he died on 21 August 1921, aged four (not five), of cerebro-spinal meningitis, according to the death certificate.

An SPR Council Member and investigator, G.W. Lambert, visited the farm and interviewed Mrs Ross and Miss Minney. Mrs Herbert had returned home to Australia. He also received a statement from Mrs Kitty Hampton, then aged ninety-five, who told him, 'I am the aunt who was keeping house at the time of his [Johnnie's] illness.... At his death he was wasted to skin and bone.'

Andrew Mackenzie, who has written (in a little more detail) a piece about this case in *Hauntings and Apparitions,* points out that it is difficult to think of a normal explanation for what took place. The percipient had only recently arrived in England and had not been in the village long enough to hear about the death of a little boy in the house where she was staying. Miss Minnie had never mentioned the subject to Mrs Ross and the only photograph of Johnnie, taken when he was three, was kept in a drawer and never seen by Mrs Ross. Critics who argue for 'coincidence' would have difficulty in accounting for the very distinctive appearance of the boy who had died there.

A crisis case

A crisis case may be defined as a recognized apparition seen, heard or felt at a time when the person represented by the apparition is undergoing some crisis.

A good example of a crisis case is the following, to be found in the SPR records. The percipient's half brother, an airman, had been shot down in France on 19 March 1917 early in the morning. She herself was in India. She explained that her brother appeared to her on the same date, 19 March, at a time when she was either sewing or talking to her baby. She cannot remember quite which but the baby was on the bed. She had a very strong feeling that she must turn round and on doing so 'I saw my brother, Eldred.... Thinking he was alive and had been sent out to India, I was simply delighted to see him, and turned round quickly to put baby into a safe place on the bed so that I could go on talking to my brother; then turned again and put my hand out to him, when I found he was not there. I thought he was only joking, so I called him and looked everywhere.... When I could not find him I became very frightened.... I felt very sick and giddy. Two weeks later I saw in the paper he was missing.'

This is a case which has often been quoted but it is typical of many others.

Post-mortem cases

These cases are defined as those in which a recognized apparition is seen, heard or felt such a long time after the death of the person represented that no coincidence with the crisis of death can be supposed. There are many cases like this; here is just one example.

Mrs P and her husband had retired to bed. Mrs P, wearing a dressing gown, was lying on the outside of the bed waiting to attend to her baby, who was in a cot. The lamp was alight; the door was locked. She says, 'I was just pulling myself into a half sitting posture against the pillows, thinking of nothing but the arrangements for the following day, when, to my great astonishment, I saw a gentleman standing at the foot of the bed, dressed as a naval officer, and with a cap on his head having a projecting peak.... The face was in shadow to me ... and the visitor was leaning upon his arms which rested on the foot rail of the bedstead. I was too astonished to be afraid, but simply wondered who it could be; and instantly touching my husband's shoulder (whose face was turned from me), I said, "Willie, who is this?" My husband turned, and, for a second or two, lay looking in intense astonishment at the intruder; then, lifting himself a little, he shouted, "What on earth are you doing here, sir?" Meanwhile the form, slowly drawing himself into an upright position, now said, in a commanding yet reproachful voice, "Willie, Willie!" I looked at my husband and saw that his face was white and agitated. As I turned towards him he sprang out of bed as though to attack the man, but stood by the bedside as if afraid, or in great perplexity, while the figure calmly and slowly moved towards the wall.... As it passed the lamp, a deep shadow fell upon the room as of a material person shutting out the light from us by his intervening body, and he disappeared, as it were, into the wall. My husband now, in a very agitated manner, caught up the lamp, and turning to me, said, "I mean to look all over the house and see where he is gone." I was by this time exceedingly agitated too, but, remembering that the door was locked, and that the mysterious visitor had not gone towards it at all, remarked, "He has not gone out by the door!" But without pausing, my husband unlocked the door, hastened out of the room, and was soon searching the whole house.'

Mrs P was wondering if the apparition could indicate that her brother, who was in the Navy, was in some trouble, when her husband came back and exclaimed, 'Oh no, it was my father!' She continues, 'My husband's father had been dead fourteen years: he had been a naval officer in his young life.' During the following weeks Mr P became very ill and then disclosed to his wife that he had been in financial difficulties and, at the time of the apparition, was inclined to take the advice of a man who would probably have ruined him.

Here again the figure is lifelike and intrudes on the percipient suddenly and unexpectedly. There is in fact no intrinsic difference between this and a crisis apparition.

Experimental cases

An experimental-type apparition is one in which someone is deliberately trying to make his or her apparition visible to a percipient and succeeds in doing so. The best-known case in the records is that of Mr S.H. Beard (well known to Sir William Barratt as a man of integrity) who wrote this report:

'On Friday December 1st 1882 at 9.30 pm I went into a room alone and sat by the fireside, and endeavoured so strongly to fix my mind upon the interior of a house at Kew ... in which resided Miss Verity and her two sisters, that I seemed to be actually in the house. During this experiment I must have fallen into a mesmeric sleep for ... I could not move my limbs.... At 10 pm I regained my normal state by an effort of the will, and ... wrote down the foregoing statements. When I went to bed on this same night, I determined that I would be in the front bedroom of the above-mentioned house at 12 midnight and remain there until I had made my spiritual presence perceptible to the inmates of that room.

Mr Beard was able to project his 'spiritual presence' into the bedroom.

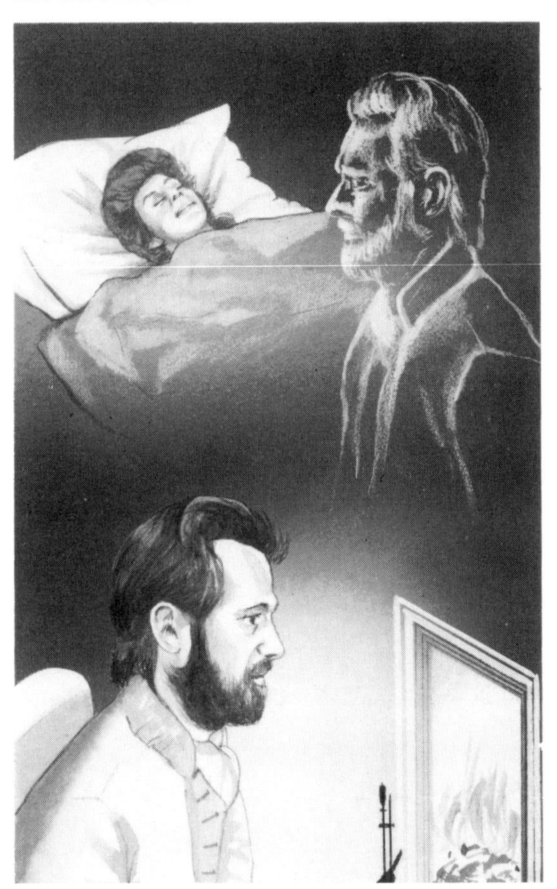

'On the next day, Saturday, I went to Kew to spend the evening and met there a married sister of Miss Verity namely Mrs L [whom he had met only once before]. In the course of conversation... she told me that on the previous night she had seen me distinctly on two occasions.... At about half-past nine she had seen me in the passage going from one room to another, and at midnight when she was wide awake she had seen me enter the bedroom... and take her hair... into my hand.... She then awoke her sister, Miss Verity, who was sleeping with her and told her about it.'

At another time Mr Beard while experimenting in this way was seen by three people.

Suggestion-type cases

Probably the large majority of the cases which one reads about in local UK newspapers (and similar media in other countries) under headlines such as 'HAUNTED COUNCIL HOUSE: FRIGHTENED FAMILY ASK LOCAL AUTHORITY FOR URGENT MOVE' fall into this category.

I was for some years the member of the SPR Council who investigated spontaneous cases. Thus it was my task to respond to letters and telephone calls to the SPR office, most of them concerned with hauntings and other ostensibly psychic phenomena. These cases cover probably ninety-five percent or so of all reported incidents but in fact I have seen very few, if any, discussions of such cases. I propose to remedy that now by a more detailed analysis of several incidents.

The factory case

One morning the SPR office was telephoned very shortly after it opened with an urgent request. It was the manager of a small cosmetics factory in Plaistow, east London, to say that his factory appeared to be haunted and, unless something could be done about it fairly quickly, his women workers would go on strike and refuse to enter the factory. They were very frightened indeed. So I went along to the factory and talked to the manager who told me of various incidents which had alarmed his women workers. As is the normal practice, I talked directly to those who had undergone the terrifying experiences. (Second-hand reports, whether hearsay or from newspapers, I have always found to be highly inaccurate.)

So the women who had had the most alarming experiences left their work of pouring mascara into little moulds or filling jars with powder. The stories they all gave to me independently agreed very well. (One might expect this as all the women talked together during their work.)

They said that some time earlier they had noticed various bangs and bumps occurring in the building, particularly on the water pipes. They also said that their radio set was not infrequently suddenly switched off and then switched on again unaccountably. In addition they showed me their store of cartons and boxes and informed me that boxes were thrown about when there was no one in the room. They all referred to footsteps walking across the ceiling (the ceiling actually being the underside of the roof space). Several of them referred to a sandy-haired man in a white coat whom they observed to walk through the factory occasionally — but not all of them could see him. They asked me whether I thought that he had been murdered in the building. An alarming experience was described to me by one of the workers who said that she was standing in the cloakroom when her skirt was pulled out to one side in an unaccountable manner.

It was clear to me that all these women were very nervous indeed and certainly were genuinely afraid to enter the factory for their daily work. After careful consideration I decided on a plan of action and the management agreed. We gathered together several of the women who had had the most frightening experiences and we held a séance after their work that evening. I searched the building to ensure that no unauthorized person was present, checking that all the doors and windows were locked. We then sat quietly around a table while the light gradually faded. Several of the women pointed out to me the sound of footsteps walking across the ceiling. However, after an hour or two, absolutely nothing untoward had happened.

The following day the manager agreed to stop work temporarily at the factory so the workers could hear the pronouncements of the SPR 'expert'.

My private conclusions were as follows: There appeared to be a normal explanation for almost all

The figure could be seen by some, but not all, of the women in the factory.

the happenings of a physical nature. The bangs on the water pipe were probably due to what is called water hammer. The system had been installed many years ago and was probably not well plumbed. There may well have been pockets of air and, in changing direction, the water would hit the pipes and sound exactly like someone striking them from outside. Water hammer is by no means uncommon. As to the radio being unaccountably switched off and on again the building had not been rewired probably for some thirty years and one could well imagine that intermittent contacts would be quite frequent. The boxes which were apparently thrown about when no one was in the store, on inspection turned out to be by no means stably stacked. There was no mystery here. They were highly likely to fall off the pile and roll on the floor. And what about the footsteps on the ceiling? The building was old and had wooden roof members. When the temperature changed, especially as it fell during the late afternoon, those long, wooden roof members would gradually shrink and, being intermittently held by friction and then moving again, would sound exactly like footsteps. This again is by no means uncommon in some buildings. Finally, regarding the skirt being pulled out: I inspected the place where it was reported to have happened and observed a considerable draught coming through a slightly open window. There appeared to be little mystery about the movement of a skirt.

So what about the sandy-haired character in a white coat? It is well known that if someone is in a state of fear and expectancy and concentrating on the possibility of something happening at any moment (that state being reinforced here by suggestions that perhaps someone had been killed in the factory) then they are likely to perceive hallucinations. This is by no means uncommon. The workers in that factory simply did not recognize the underlying reasons for any of the phenomena which they had observed. These phenomena appeared to have led to the full-scale hallucinatory experiences of the more nervous and suggestible of the women. So what to do?

I addressed all the women workers, who waited with bated breath to hear my pronouncements. It would have been of no use for me to talk about the dramatizing capabilities of the unconscious mind and how the state of fear and expectancy resulting from the misunderstood normal phenomena was producing a hallucinatory, sandy-haired white-coated man. The women would merely have said 'Nonsense, he doesn't know what he's talking about; we could see the ghost perfectly clearly!' Which indeed they could! It was necessary instead for me to invent something which they would find acceptable and which would remove the suggestions that had occurred. So I explained that in order for physical happenings to occur, energy was necessary. I had often observed that when an investigator writes down notes concerning exactly what has happened at a haunting this appeared to discharge the energy and the phenomena usually cease. The séance the previous evening, during which nothing unusual had happened, showed very clearly that all the energy had been discharged. (I observed a great sigh of relief from all the women.) So I said very positively: 'There will be no further happenings. You can all relax: it is most unlikely that anything more will occur'.

The following afternoon I telephoned the manager. There had been no more sightings of the sandy-haired character even though up to that day there had been several daily. I telephoned again a week later. Still no phenomena! And again a month later! The tail-piece to this story was that the cosmetics factory presented my wife with a box of cosmetics and made a small donation to the funds of the SPR!

The haloed figure.

The cosmetics haunting did not figure in the local newspapers. But this next one did. It was, if I recollect accurately, a reporter who asked for an SPR investigator. I visited a small terraced house in London and discovered the lady of the house in a state of fear and unable to go alone into her sitting room. She agreed to go in there with me and we discussed her alarming experiences. She explained that frequently the electric cord joining the television to the wall socket unaccountably swung to and fro. She also told me that she could sometimes see a figure outlined in a phosphorescent glow peering at her through the frosted glass separating the sitting room from the kitchenette. The neighbours had informed her that they thought someone had been murdered in that room! The latest terrifying experience had been her sighting of a shadowy figure which walked out of the kitchenette, across the sitting room and then disappeared through the wall.

While I was sitting talking to her I noticed considerable vibration from trucks which were passing along a nearby road. There was little mystery about the swinging of the television lead. I then went into the kitchenette at the other side of the frosted glass and looked through the window which was directly opposite at the far end. I could see at the other side of a piece of ground a street lamp. At night this would have been brightly lit and when seen through the frosted glass would appear like a figure with a halo round it. (Imagination can play strange tricks.) The woman who had experienced the hallucinatory figure was certainly in a state of fear and expectancy; she really thought she was being besieged by creatures from the next world!

Seen through the frosted glass, the street lamp became a frightening haloed figure.

The story necessary to 'exorcise' this ghost was very similar to the last one. I explained again that writing down all the details often seems to discharge all the energy and that frequently the visit of an investigator leads to the cessation of all phenomena. I told her very positively several times that she could relax as nothing further would occur.

Checking later, the result was as expected. The resolution of her anxiety and stress completely removed the phenomena.

Shadow creatures

I could continue with many stories like this. However, spontaneous hauntings are not always similar. Just occasionally they turn out differently. I remember one case of a family who lived in a very old house in Essex who telephoned the SPR for advice and help. They had some months before returned to England after many years spent in India. They had bought a very old house and the wife was observing shadowy figures crossing the passage and shadowy animals leaping around the bedroom. The husband, a rather down-to-earth military man, had no experiences of that kind but he did say that he felt some sort of atmosphere and was anxious about his wife.

A five-minute chat over tea elicited that the wife frequently had psychic experiences. When I asked her whether she sometimes knew beforehand when a friend was about to call, physically or by telephone, she said excitedly that she did and that it happened not infrequently. She also described other experiences of a similar nature which showed me very clearly that she was psychic. I asked her whether she knew what being psychic was. She said that she had no knowledge of that as she was a Catholic and a priest had told her many years before to 'keep away from Spiritualism'. (Something must have happened to her at that time to promote such advice.) So I told her that she was herself psychic, that she was very fortunate in that she had a window on the universe not vouchsafed to a high proportion of the population and that there was nothing to worry about. Psychic people did occasionally see shadowy figures walking around. Again, all the stress disappeared from her face and she became more relaxed. A week or two later I telephoned and she said that most of the phenomena appeared to have died away. Just occasionally she had faint intimations of something but it was no problem to her. In such cases the state of stress seems to lead to more positive phenomena and its removal appears to reduce their intensity.

Shadowy animals appeared to leap around the bedroom.

COMMENTS

What conclusions can be drawn from these examples of the five classes of apparition? First, there should now be an appreciation of how very disparate the various types are; there is not just one classical haunting-type ghost. Secondly, the many similarities between the various kinds of experience should also be clear. The apparition is in most cases so like a human being that it is at first mistaken for one. Sometimes it is rather shadowy and evanescent. The figures therefore are very like normal human beings but sometimes behave as though sleepwalking (the haunting-type apparition is more like a sleepwalker than are the other types).

However, let us start the discussion by following the example of an earlier writer on this subject, G.N.M. Tyrrell, and sketch a 'perfect apparition' showing at once all the characteristics of apparitions which are known with reasonable certainty. If a perfect apparition were standing next to a normal human being we should see practically no difference between the two. The apparition would probably behave as if conscious of us. It might even touch us. If this occurred or if we came near to it we should probably feel a sensation of cold. If we spoke to the apparition we might receive an answer but could have no long conversation with it. It might open and close a door or pick up an object. We should both see and hear those objects move yet the sight and sound would not be recorded on film or tape recorder. If we attempted to grasp the apparition our hand would go through it without encountering resistance. If the apparition walked across a room where chalk dust had been spread on the floor and threads stretched across its path it would walk through without any physical mark or disturbance. After a certain time – it might be up to half an hour or so – the apparition would disappear by suddenly vanishing or by becoming gradually transparent and fading away. It might vanish through the wall or the floor or it might open the door and walk out. Some apparitions may be slightly self-luminous.

That is what an apparition would be like at its best according to reliable evidence; it is, Tyrrell says, what has through the ages been called a spirit. It is clearly in reality a psychological phenomenon.

Post-hypnotic suggestion

Let us now briefly consider the phenomenon of post-hypnotic suggestion because it is relevant. Good subjects can be hypnotized and given a post-hypnotic suggestion that someone will come into the room, walk round it and go out again through the door. They can then be awakened and restored to normality. However, they will then observe someone come into the room, walk round it and go out again. They will observe no difference between that person and any other person present normally. They might even engage the new person, the result of the forgotten suggestion under hypnosis, in conversation and receive answers to questions (on checking these will probably be fictitious). And if that conversation is recorded with a video camera, only they themselves and their side of the conversation will appear. No one else in the room would be able to see the new 'person'.

Remember the 'iceberg concept' of the mind (see page 11). It is quite clear that at some level of a subject's unconscious mind the suggestion of the hypnotist that a person would walk into the room has been taken and accepted. That part of the mind was called 'George' in chapter 1. So George accepted the suggestion and dramatized a character to suit, dressing him or her in appropriate clothes, equipping this person with a personal history and blank-

A suggestion given under hypnosis will later be 'seen' and accepted as a normal event.

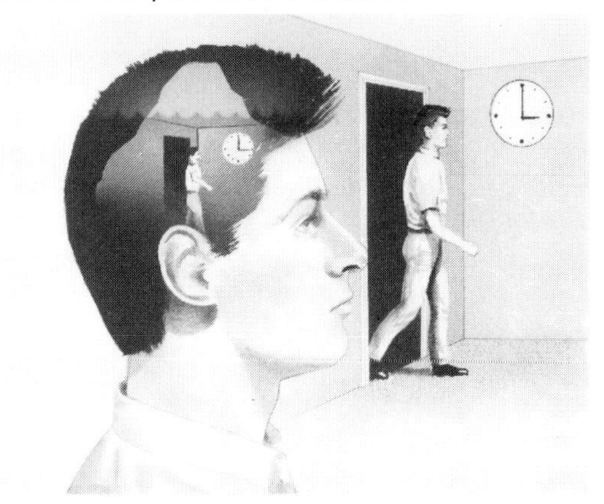

ing out the necessary parts of the background where the character was supposedly standing.

Let us now apply this knowledge of the unconscious mind and hypnotism to help us understand apparitions. The cases of apparitions (like hypnosis) reveal the existence of a machinery in the personality which can produce sensory imagery — imagery that is just like normal perception and quite as fully detailed. The next question which naturally arises in our minds is, 'Who supplies the suggestion?'

The answer to this question may not always be so simple as it appears. However, in the case of a crisis case a first choice would obviously be the person experiencing the crisis. This would be an example of telepathy, the suggestion being given to George telepathically. In the case of an experimental apparition, clearly a first choice would be the person carrying out the experiment. In a post-mortem case the deceased person perceived might be a first choice provided one were reasonably convinced of human survival beyond bodily death (discussed in more detail a little later in the book). The appropriate suggestions in the case of a suggestion-type apparition could be from a number of sources — the misunderstood phenomena, the fears of the neighbours and so on.

The haunting-type case is perhaps a little more difficult but one would imagine that the person perceived as the apparition perhaps originally had something to do with it. Professor H.H. Price suggested a sort of enduring 'psychic charge', perhaps the results of strong emotion, may be present in the fabric of the building. In haunting cases it is very much as though a piece of photographic film is being run through an unconscious projector (as mentioned earlier). Clearly slight variations are possible as an apparition sometimes appears to take notice of an observer, even though speech is exceedingly rare.

It will be perfectly clear that this is by no means the whole story. The simple answer is that we do not really know what causes apparitions. Let us illustrate this with the post-mortem case described earlier — where the naval officer was suddenly observed leaning on the bedrail. It could have been that the husband of the first percipient was only too well aware of the possibilities that the advice he was about to take concerning his financial affairs was unwise and that if his father had been present he might have warned him of this. Perhaps George produced the apparition of his father in order to make this perhaps partially unconscious feeling overt. However, in my view the evidence for survival is exceedingly good and we can by no means rule out the possibility that surviving relatives are sometimes aware of our activities here on earth and might be able to find some means of communicating with and influencing us.

EXORCISM

Quite often people who are troubled by an apparition call in a priest rather than a parapsychologist. The priest perhaps refers to another priest who is an expert in 'exorcism' and eventually a service of exorcism, involving forms of words and the sprinkling of holy water, is conducted at the site of the apparition. Sometimes this is effective; sometimes the apparition does not appear to know what is expected of it and continues as before. I have occasionally suggested that my own psychological methods of removing a suggestion-type apparition (which is a high proportion of the total) have been more effective because I have never had a failure. However, it may be that, in the case of earnestly believing Christians, the service of exorcism will act as a very strong suggestion to counteract whatever caused the apparition and is very effective in removing it. (This, by the way, is quite a different theory from that which would have been advanced by the priest conducting the exorcism.) The attitude of those involved is all important, whether with regard to exorcism, psychic healing or séances; the enormous importance of belief is paramount.

MY OWN APPARITION

I cannot finish this chapter without including my own apparition! (I am numbered in the one in ten of the population who have had such an experience.) Some years ago I gave a lecture in a country town in the Midlands and afterwards was shown to a room in a guest house where I was to sleep. The building was, if I recollect correctly, Tudor. My room was small and the bed was surmounted by the underside of the steeply sloping roof. It was late and I at once retired for the night, the moon shining through the window beyond the foot of the bed. After some time (I have no idea how long) I awoke, well-aware of where I was, and observed the moon still shining outside the window. To my surprise I now noticed that the ceiling over my face appeared to have in it a ragged hole through which another face was gazing down at me. This face seemed to be illuminated by a flickering candle, which I could not see, and had bright, large unmoving eyes, looking straight into mine. 'Well', I thought, 'What a cheek, looking at me in bed in the middle of the night!' Then I fell off to sleep.

In the morning, after a very early and solitary breakfast, I had to leave and had no opportunity of making enquiries for some months. The organizer of my lecture then informed me that the building in which I had slept was reputed to be haunted, but had no details.

I would blame no one for suggesting that I had dreamed the whole episode. However, all I could say in response would be that I have never had a dream anything like that one at any other time. An interesting feature was that I was clearly not in a fully normal and conscious state as it did not seem to occur to me to attempt to grasp the entity gazing so fixedly at me. If I had been fully conscious rather than slightly dissociated I doubt whether I should have had the experience. It is rather rare for someone lying in wait to perceive an apparition.

Before leaving this subject it is necessary to put forward one further theory of apparitions. It has been shown that, according to the telepathic theory, apparitions are not to be considered as objects in physical space, even though they behave like it. They may, however, quite sensibly be thought of as objects in another space, other than the physical. An alternative theory considers apparitions as objects made of 'psychic ether' perceived when psychically illuminated. Yet another theory, not too different from the latter, will be considered in the chapter on the séance. None the less there is still plenty of scope for original thought!

In chapter 3 we shall examine experiences that may be more clearly recognisable as psychic.

3
TELEPATHY AND CLAIRVOYANCE

EXTRASENSORY PERCEPTION: ESP

Extrasensory perception, which is sometimes called paranormal cognition, is defined as the acquisition of information otherwise than through the recognized sensory channels. The information may be of facts relating to another place or of thoughts in someone else's mind; the information acquired can lie in the present, past or future.

Telepathy is the communication of impressions of any kind from one mind to another mind, independently of the recognized channels of sense. (Precognitive telepathy is the similar acquisition of information concerning the state of another mind but which lies in the future.) Clairvoyance is defined as the paranormal acquisition of information concerning a physical event or an object elsewhere, the information being derived (it is assumed) directly rather than through the mind of some other person. Precognitive clairvoyance would be the similar acquisition of information lying in the future. Precognition is defined as knowledge of some future event which cannot be inferred from present data. As will be seen, there is frequently doubt whether information within someone's mind, not acquired through the senses, is the result of telepathy or clairvoyance. It is necessary to carry out some rather specific experimental procedures in order to separate the two.

Let us have a few examples so that we have a clear idea of the sort of thing we shall be discussing. If a subject has an experience of observing a crisis apparition – for instance a soldier in uniform dying, and so acquires the information that her son has been killed at about that time, and if the confirmation comes only later by telegram – then this would be an example of ESP. It would probably be of the kind described as telepathy from the son, the information being made overt (pushed up into the conscious mind) by George in the form of an hallucination of the dying son (see also page 17).

Again, if the subject dreams that her son is dying, awakening with a strong feeling of the dream's reality, this may be the way selected by George to make the unconsciously acquired information overt, the conscious mind being quiescent at the time and perhaps easier to impress than when it is in use during the day. Of course the sceptic is entitled to say that either experience may be the result of a mother's deep concern for her son's safety. If this is the case then there will be no information a little later of the son's death or, the sceptic would say if there were such information, that it was the result of chance coincidence. Another not uncommon example of ESP is a mother's waking up with a strong feeling of concern for her child, then going to the child and finding it in a place of danger. The sceptic can make the same comment about this. It is true of course that there are lots of examples of coincidences; many people have worries about their friends and relatives, most of which are not borne out in reality.

How does one eliminate such chance coincidences and find out whether ESP (there are several kinds) really does occur? The answer is of course by properly controlled laboratory experiments to eliminate chance coincidence and the normal transfer of information. But before we consider these it will be useful to examine some more examples of how, in the theory we are investigating, George manages to push up into the conscious mind the unconsciously acquired information. He has a number of ways of doing this – crystal gazing, casting a pile of bones on the ground, observing the entrails of animals, seeing pictures in tea leaves, and so on. In the case of some psychic people, George can produce very satisfactory pictures using a sphere of glass (a so-called crystal) as an appropriate matrix. Other psychics can sometimes observe pictures which they describe as seeming to be projected on the wall or in space in front of them. These pictures sometimes have symbolic components which the psychic has to learn to interpret. The unconscious mind, as is very well known to psychologists, frequently works in symbols and such symbols are often to be found in dreams. George, it will be recalled, has no conception of logic as does the conscious mind. Sometimes he associates two quite disparate experiences and permanently links them together.

Other ways which can be used by George to transfer information from the unconscious to the conscious mind are the divining rod (used by water or other diviners — who might also use a pendulum) and so-called automatic writing. In both these cases George directly controls the muscular apparatus without the conscious mind being involved. The same method is used when words are spelled out using a Ouija board or its common-or-garden equivalent when the letters of the alphabet and the words YES and NO are written on pieces of paper and put in a circle round the table; an upturned glass is used to slide to, and so indicate, the letters, the sitters each placing a finger gently on the glass and resolving not to push. The information which George brings to the conscious mind by one or other of these methods is not of course always accurate: it may be fantasy.

The question is often asked, 'Is ESP a throwback to an earlier state of evolution before the five senses were so well developed — or is it an incipient all-embracing sense due to be unfolded after further evolutionary development?' The honest answer is that we do not know. For what it is worth, I would guess that it may well be developed by humanity in general in the future. However, as we shall see later, it is probably a faculty which is, as it were, squeezing through from other 'levels of consciousness'. The near-death experiences that will be discussed in chapter 6 should provide us with some interesting facts for consideration.

EXPERIMENTS

How can we conduct experiments which distinguish between telepathy and clairvoyance? Let us first have a reminder of the difference. Take for example a card-guessing experiment. An 'agent' is looking at cards selected with the guidance of a random-number table; a 'percipient' is stating what he or she guesses those cards to be on a buzzer signal from the experimenter who is with the agent. In this instance the cards might be guessed correctly

'Seeing' a card directly implies clairvoyance: obtaining the information from someone's mind is telepathy.

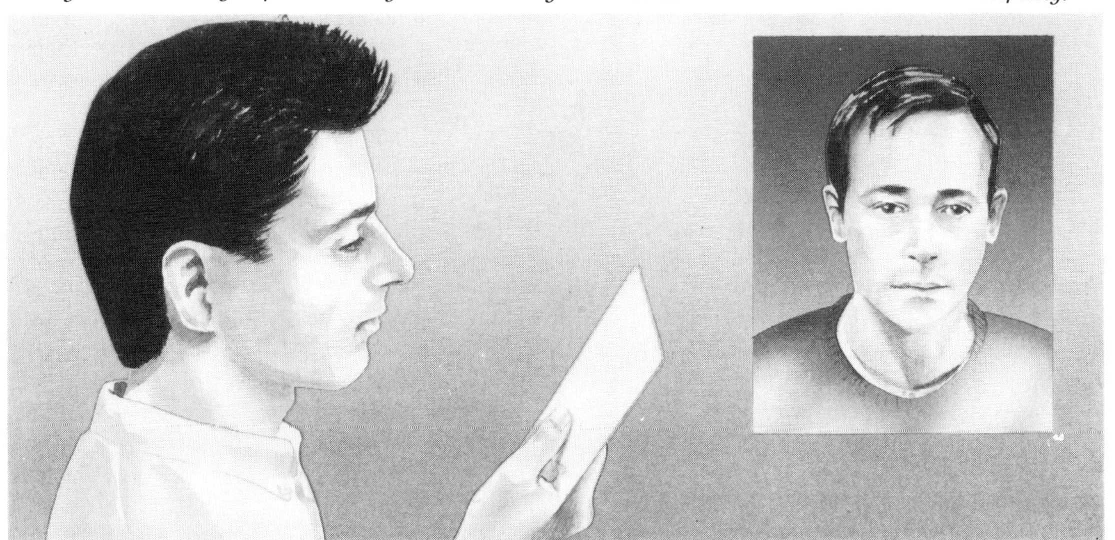

either by telepathy from the agent's mind or by somehow directly perceiving the cards, the agent acting only as a sort of beacon, if in fact an agent is necessary at all. There is no way of telling to which of the two faculties success should be attributed and such experiments are referred to as testing 'general extra-sensory perception'.

It has sometimes been suggested that there is no way of distinguishing between telepathy and clairvoyance because someone must always know the answer, if not coincidentally then later. But this is not at all true. It is by no means difficult to separate telepathy and clairvoyance — as the following example will show.

The random number box

One of my electrical engineering undergraduate students designed and built for his final-year project a box which would carry out the following procedures: If a button were pressed at the front of the box then a three-digit random number would appear at the back. If the user of the box thought that he or she knew what the number at the back was (by any means at all) then he could set three dials at the front of the box to those digits — and then press the button again to obtain a new random number at the back. The first random number would be lost but if the numbers at the front and the back of the box were in agreement then a count of 1 would be scored on a 'correct' counter within the box. If the numbers disagreed then a score of 1 would be recorded on an 'incorrect' counter. The aim of the experiment would be for a psychic to state by ESP what the numbers at the back were on each occasion without anyone looking at the numbers in the ordinary way. At the end of a run of 'guesses' the dials at the front of the box could be set to indicate the total number of rights and wrongs (agreements or disagreements).

It will be appreciated that in this experiment no one at any time looks at the numbers at the back and they are never known to any living mind through the channels of the senses. If the psychic states correctly any of the numbers at the back then it is possible to calculate the mathematical odds against chance as accounting for the agreements. If these odds against chance are very high then it is reasonable to say that the psychic has a paranormal knowledge of a proportion of the numbers. This experiment completely eliminates any telepathy, including pre-cognitive telepathy — in which it is imagined that the psychic might obtain information regarding the state of someone's mind which, so far as the ordinary consciousness is concerned, lies in the future. (The evidence that this does sometimes occur is very good, as is the evidence that psychics can sometimes obtain information which was at some time in the past in someone's mind, called retrocognitive telepathy.)

Obviously, if a psychic prefers geometrical symbols rather than numbers then the box can easily be arranged for these instead. In addition, if a psychic prefers living things like plants or fruit, these can, by a little elaboration, be presented in small boxes out of sight of the psychic. The same considerations apply to all these experiments. Success has been obtained in many experiments of this kind. However, I shall now give two examples of how confusing George can sometimes make the picture when one carries out an experiment like this.

Ingo Swann, a distinguished American psychic from New York City, once came to visit me in the University. He arrived without much warning. I told him of the random-number box and he immediately expressed a wish to try a run. I explained that the correct working of the box had not been recently checked (it had been in my cupboard for some months) but none the less he wished to have a go. I gave him about twenty guesses and at the end of the series when I set the dials to indicate how many of the three digit random numbers he had guessed correctly the box indicated eight. This was a quite astonishingly high number and there would have been enormous odds against chance accounting for it. Assuming all was well, I made arrangements to carry out further experiments with him by the transatlantic telephone after he had returned home. The following morning I switched the box on again and tried a run of my own. I also achieved eight 'correct'! Something was clearly wrong. My research assistant had a look at the circuitry and discovered that certain bad contacts were causing all the seven bars on the light-emitting diodes — which were illuminated to form the different numbers — to be illuminated when the dials were set for the total score of agreements. He cleaned up the contacts, restored the microchip, and all was back to normal: I scored zero unless I actually looked at the numbers at the back and set the numbers in the front to match these.

An opportunity occurred later to do the experiment with a well-known British psychic, Matthew Manning. This time there was plenty of notice and the equipment was switched on early in the morning so that it reached a stable temperature. Also I ran a series of trials, sometimes looking at the numbers and arranging an agreement and sometimes not.

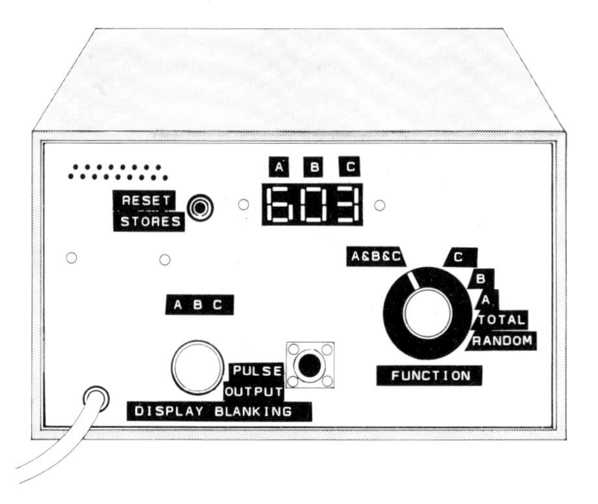

The random-number box enabled Professor Ellison to eliminate telepathy when testing psychics and led to some very interesting and surprising results.

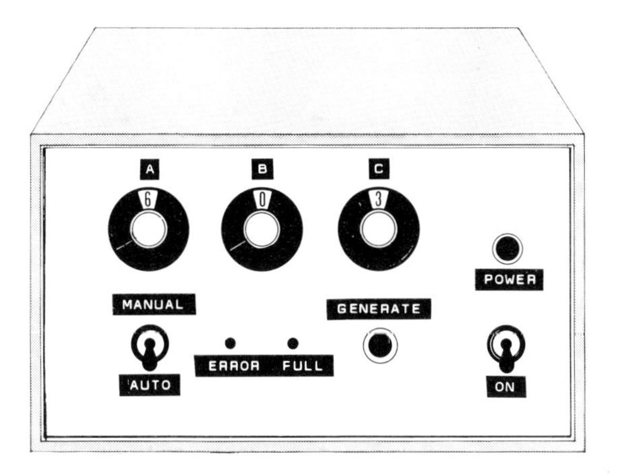

The box gave the correct number of agreements at the end. My research assistant did exactly the same and everything worked perfectly. When the psychic arrived we did a run of twenty pure clairvoyance experiments with him and turned with interest to see how many he had scored. The box read eight again! We cleaned up the contacts once more, replaced the chip and the research assistant and I did runs to check that all was working perfectly. The psychic did another run and again he scored, according to the box, eight correct!

What does one deduce from an experiment like that? The first thing is that it would not convince a sceptic that the psychic had psychic faculty. The sceptic would merely say that by chance the anomalous behaviour had occurred when the psychic was doing the experiment. However, what can a psychical researcher reasonably consider might have happened? Perhaps George has discovered that using 'pure clairvoyance' to see the numbers at the back of the box is, in those circumstances described, very difficult. However, in his a-logical way he knows that the experimenter wishes to achieve a high number of 'agreements': that the goal is a high number. He finds that it is much easier to use psychokinesis to spoil the contacts. This may have been what was done in each case — but there is no way of finding out for certain!

As will be seen in later chapters, the operation of psychic faculty is 'goal directed': it is not necessary for there to be a conscious understanding of the circuitry or mechanism to achieve the final desired result. How this can be we have no idea! But the evidence that it is so is strong.

An experiment on telepathy rather than clairvoyance is rather easier and would involve an agent's choosing mentally a target which the percipient would attempt to guess. Provided nothing is written down then success would appear to be due to telepathy rather than clairvoyance — but there is no means of proving that there was not some sort of way, which is exceedingly difficult to follow conceptually, that the percipient discovered the physiological (electro-chemical) state of the brain of the agent, rather than the concomitant thoughts, and by this route discovered what the target was supposed to be. This is a difficult and rather elusive subject to examine!

J.W. Dunne and dreaming

A well-known experimenter of some years ago was J.W. Dunne who conducted experiments in dreaming. By writing down the details of his dreams Dunne appeared to be able to show that some of them were precognitive — that they foretold the future. There has been controversy about this, centering particularly around chance coincidence and certainly his ideas of time are not these days taken very seriously. (J.B. Priestley's 'time plays' were written after he had read Dunne's ideas.) It seemed to me that it might be worth while trying that experiment again and so for about a year I went to bed with a pad and pencil beside me, resolving to write down my dreams and see whether they were ever precognitive. What was the result? George quickly responded to the idea that I wished to have dream information and he produced so many dreams, waking me up immediately after each one, that I had hardly any sleep. I wrote them down rigorously but did not seem to see any very definite evidence for precognition. However, I was losing so much sleep I had to abandon the experiment! (In similar ways George will collaborate with a psychiatrist and produce the appropriate symbols to give information consistent with the theory upon which he or she is currently working.)

Telepathy between children

It has been suggested that young children are more likely to possess ESP than adults because adults, especially here in the West, have been conditioned by their culture to consider that such a thing is impossible. If this idea is true and telepathy does take place, as indicated by other evidence, then it should be stronger between young children and will decline with age. This has indeed been shown to be the case. Spinelli found that children aged three to eight years scored positively in a test for general extra-sensory perception, with the youngest groups scoring highest.

DISTANT VIEWING

Drs Russell Targ and Harold Puthoff at Stanford Research Institute devised some experiments to study what they called 'distant viewing'. Oversimplifying somewhat, the experiment was as follows. One of the two experimenters would sit down with a subject (or 'percipient') in a comfortable room and lock the door. The other experimenter would then go to an independent person and ask for an envelope to be selected at random from a number of other sealed envelopes. All the envelopes contained details of sites within about fifteen minutes' drive of SRI. The second experimenter would receive the random envelope, open it, and drive to the place specified therein. At an agreed time (the door of the room containing the first experimenter and the subject had been locked) the second experimenter would wander around the randomly chosen site, looking especially at the features of the site shown in the photographs in the envelope and also described. At that time the first experimenter would be explaining to the percipient that he or she should relax and state whatever impressions, no matter how fleeting, floated into the mind. The result would be descriptions and sketches by the percipient in one sealed, labelled envelope; another would contain actual details of the target site.

The experiment would be repeated exactly as described perhaps twenty times. The procedure then would be to hand the twenty targets, perhaps numbered one to twenty together with the twenty descriptions perhaps lettered (in random order obviously) A to T, and for an independent jury to rate the matches between each target and every description of the agent. They would rate the degree of correspondence perhaps on a scale of 0 to 5, a really good agreement in a lot of features being 5, and no agreement at all being 0. The result of this could be a mathematical analysis showing how well the different sets of perceptions of the percipient agreed with the various targets. Targ and Puthoff sometimes had complete identification of a target. The agreements ranged from such an absolute identification, including naming, to a complete lack of any common features.

Robert Jahn, the Dean of Science and Engineering at Princeton University, together with teams of

Target: Ruins of Urquardt Castle, Loch Ness, Scotland. The percipient was in New York City, some 3500 miles away. The perception, generated 14 hours retrocognitively, reads:
"Rocks with uneven holes. Also smoothness. Height. Ocean. Dark. Dark blue. White caps. Waves booming against rocks? On mountain or high rocks overlooking water. Dark green in distance. Gulls flying? Pelican on a post. Sand. A lighthouse? Tall structure. Round with a conical roof. High windows or window space with a path leading up to it. Or a larger structure or a castle." (Here there is a sketch of a castle abutment on the transcript.) "Old. Unused. Fallen apart. Feeling musty, or dark. Moss or grass growing in walls. Wood draw bridge? A black dog? Snow. Ice-capping a mountain. High large cavernous hall. Castle. Strong positive emotional response."

An example of a good distant-viewing experiment.

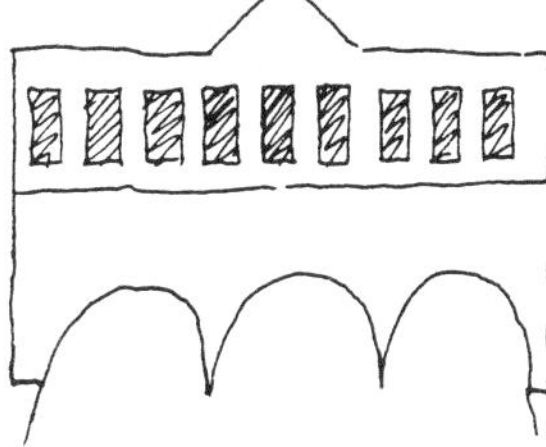

One of the most interesting points to emerge from distant-viewing experiments is that success does not vary with the time factor: a good 'percipient' is just as likely to succeed precognitively — that is, before the site has been randomly chosen.

electronics and computer engineers and statisticians, greatly improved the protocol for this type of distance viewing experiment — which Jahn calls 'remote perception'.

In Jahn's language, the percipient attempts to visualize or sense the details of a geographical target physically remote and inaccessible through ordinary sensory channels, using any strategies of consciousness he or she finds effective. He states that the target is usually 'defined' by an agent directed there at a specified time via some random procedure. (The word defined is used as it is of course not clear whether telepathy or clairvoyance is being used.) In Jahn's version of the experiment, data are accumulated and processed in binary form. The agent photographs the target and writes a description of it, but also ticks off answers to thirty binary questions serving to define the target in a particular code (is it indoors or outdoors? noisy or quiet? are there cars or no cars? is there water present or not? and so on). The percipient also ticks off answers on a similar list, in addition to producing a description with sketches. All scoring procedures are based on these two sets of binary responses. Jahn's team has explored fifteen or twenty scoring methods to score the matches between the targets and perceptions.

In its powerful way the computer can compare any perception with all three hundred targets in the data base. The mismatch scores for any given method are found to define what Jahn calls an 'empirical chance distribution' that is remarkably Gaussian (the well-known bell curve of the 'normal' distribution). This enables a quantitative statistical evaluation of the matched scores. In Jahn's words the mismatched scores are 'the off-diagonal matrix elements of the perception/target array' — the 'empirical chance compounds' that permit the statistical analysis. Jahn subtracts the empirical chance distribution from the results and gets a residue — which in one example was about fifteen per cent of the data. The possibility of achieving this residue by chance was one in ten to the power twelve (in other words, there was one possibility in a million million of obtaining that residue by chance) and this was much the same odds as given by the other scoring methods tried.

The natural question for the physical scientist to ask next is, 'How does this extra-chance component depend on the distance between the agent and the percipient?' Jahn's team did a large number of tests at different distances between agent and percipient and found no discernible preference for nearer targets. Secondly, one would naturally wonder about the variations of the score with the time interval between the effort of the percipient and the visit to the target. The Princeton team found no discernible dependence of the perception accuracy on the time interval between perception effort and target visit and that variation in time interval was made not only positive but negative. In other words, percipients had their perceptions not only for various times after the target had been visited by the agent but also for various times beforehand – before the target had even been selected let alone visited. Because most data are acquired precognitively – the perceptions are acquired and the data sheet completed hours or days before the target is visited or, in most cases, before it is even selected – Jahn calls this experiment 'precognitive remote perception'. As Jahn says, 'How the consciousness of a percipient can get access to points remote in time and space from the current location is well beyond our physical understanding'.

Jahn has a programme trying to produce various kinds of model to represent all the parapsychological facts they are acquiring. They are looking at 'electromagnetic models, thermodynamic and mechanical models, statistical mechanical models, hyperspace models, quantum mechanical models and others'. Jahn feels that a more fundamental level of modelling than all this will need to be used. He suggests that we have to reconsider what we mean by 'reality', 'consciousness' and by 'environment'. For the moment let it suffice to say that in our iceberg model of the mind George has, through levels of the unconscious further down in that model, access sometimes to information which is not available to the ordinary physical consciousness by reasons of time or space.

ANALYSING THE EXPERIMENT

The *30 Binary Questions* shown on the next page are typical of those used by Robert Jahn to analyse distant-viewing experiments. For statistical purposes, a great number of experiments would be needed before the results could be regarded as scientifically relevant. Moreover, the 'weighting' of the questions shown here is variable. (For example, more scenes tend to be outdoors than indoors, more tend to have people present, fewer have animals, and so on.) However, using these questions will help with a clear assessment of any such experiment. The results could well prove interesting. (For a single simple experiment it may be better to use the method described on page 39.) The 'descriptors', thirty in number, are posed in binary form, to be answered yes or no. This will enable them to be fed into a computer for analysis. They range from factual aspects – such as whether the scene is indoors or outdoors, or whether there are trees or vehicles present – to the more subjective aspects such as whether the area is noisy or quiet, confined or open, busy or tranquil. The list should be given to the participants in this order.

If pure chance only is present the line will wander near the horizontal, having 1 chance in about 20 experiments of crossing the dotted line. The further the line deviates from the horizontal, the greater the probability that the result is not due to chance. It is seen to exceed the 1 in 20 chance by an increasing margin the larger the number of experiments showing that remote perception is taking place.

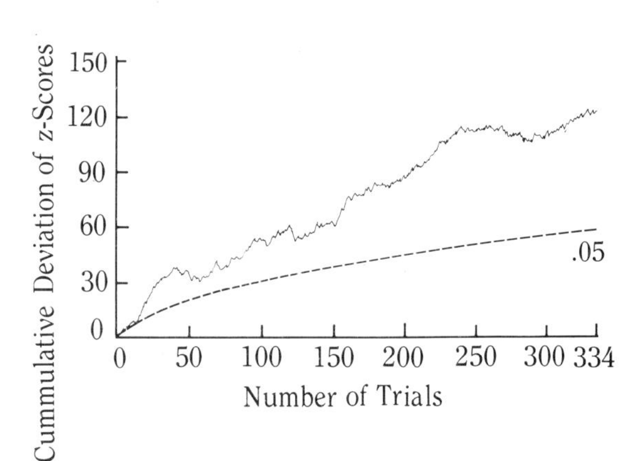

30 BINARY QUESTIONS

1. Is any significant part of the perceived scene indoors?

2. Is the scene predominantly dark; for example, poorly lit indoors or night time outside?

3. Does any significant part of the scene involve perception of height or depth, such as looking up at a tower, tall building, mountain, vaulted ceiling or unusually tall trees, or looking down into a valley or down from any elevated position?

4. From the agent's perspective, is the scene well-bounded, such as the interior of a room, a stadium or a courtyard?

5. Is any significant part of the scene oppressively confined?

6. Is any significant part of the scene hectic, chaotic, congested or cluttered?

7. Is the scene predominantly colourful, characterized by a profusion of colour, or are there outstanding brightly coloured objects prominent such as flowers or stained-glass windows?

8. Are any signs, billboards, posters or pictorial representations prominent in the scene?

9. Is there any significant movement or motion integral to the scene, such as a stream of moving vehicles, walking or running people, or blowing objects?

10. Is there any explicit and significant sound, such as an auto horn, voices, bird calls, or surf noises?

11. Are any people or figures of people significant in the scene, other than the agent or those implicit in buildings or vehicles?

12. Are any animals, birds, fish, or major insects (or figures of these) significant in the scene?

13. Does a single major object or structure dominate the scene?

14. Is the central focus of the scene predominantly natural, that is, not man-made?

15. Is the immediately surrounding environment of the scene predominantly natural, that is not man-made?

16. Are any monuments, sculptures or major ornaments prominent in the scene?

17. Are explicit geometric shapes — for example, triangles, circles or portions of circles (such as arches), or spheres or portions of spheres (but excluding normal rectangle buildings, doors, windows, and so forth) — significant in the scene?

18. Are there any posts, poles or similar thin objects, such as columns, lamp posts or smokestacks (excluding trees)?

19. Are doors, gates or entrances significant in the scene (excluding vehicles)?

20. Are windows or glass significant in the scene (excluding vehicles)?

21. Are any fences, gates, railings, dividers or scaffolding prominent in the scene?

22. Are steps or stairs prominent (excluding curbs)?

23. Is there regular repetition of some object or shape, for example, a lot full of cars, marina with boats or row of arches?

24. Are there any planes, boats or trains (or figures thereof) apparent in the scene, moving or stationary?

25. Is there any major equipment in the scene, such as tractors, carts or gasoline pumps?

26. Are there any autos, buses, trucks, bikes or motorcycles, or figures thereof, prominent in the scene, moving or stationary (excluding agent's car)?

27. Does grass, moss or similar ground cover compose a significant part of the surface?

28. Does any central part of the scene contain a road, street, path, bridge, tunnel, railroad tracks, or hallway?

29. Is water a significant part of the scene?

30. Are trees, bushes, or major potted plants apparent in the scene?

Reproduced by kind permission of Robert Jahn

GOVERNMENTAL SUPPORT FOR RESEARCH

It is evident that the Government of the United States of America has provided support for research on parapsychology: there can be no doubt about it. Certainly contracts have been placed with Stanford Research Institute. The amounts, compared with those put into research in other defence areas, are very small but nonetheless they exist. Information regarding American Government interest is to be found in *The Mind Race* by Targ and Harary where a description is given of the Government's sponsored research programmes in both the USA and the USSR. Fair-sized sums have undoubtedly been spent in successful efforts to develop useful psychic functioning. The 1980 report of the Congressional Committee on Science and Technology referred to the successful results of parapsychology, especially regarding SRI's 'remote viewing'; it mentioned the apparent 'interconnectedness' of minds and matter and referred to the possibilities of obtaining information independent of geography and time. It is clear to anyone that the potentiality of this knowledge is powerful and far-reaching and the report mentions the high level of official support for the research in the USSR and suggests that Congress might undertake a serious assessment of their own research in the United States.

It seems clear that the USSR has officially sponsored parapsychological research for over sixty years but reports on this research have come usually from journalists unable to assess it properly. Targ and Puthoff, assisted by Larissa Vilenskaya who was active in Soviet parapsychological research for more than ten years, suggest that much of the research in the Soviet Union has been in attempts to modify the behaviour and feelings of remote humans and animals by psychic means. Very early Russian work was involved with the control of behaviour at a distance by hypnosis.

POLICE AND TELEPATHY

Another official use of psychic functioning, in particular distant viewing, has been its employment by the police of various countries. The Dutch psychic, Gerard Croiset, was often telephoned by police departments from other parts of the world to ask him to see whether he could obtain impressions regarding the likely whereabouts of missing persons. He once described to me how, as he talked on the telephone (perhaps to a police commissioner as far away as Canada) impressions of geographical features of a particular environment would float into his mind and these he would describe. Such features were often of value to the police in discovering hidden bodies following murders.

It is obviously particularly difficult to obtain exact details of these cases because the police authorities are naturally shy about admitting that they attach weight to this sort of thing, especially in the presence of certain sceptical scientists. However, I well remember several years ago in New York City watching a car set out carrying a policeman and a psychic (Dr Alex Tanous) whom I knew well, the aim being to see whether he could spot in the New York crowds a certain murderer who was at large. It must have been like looking for a needle in a haystack and he was not successful on this occasion. The biographies of several British psychics refer to police use of their capabilities but again a certain vagueness is evident. Anyone who has read the biographies of psychics will perhaps have noticed or been suspicious of the almost uniform success they describe. Of course real psychic functioning is not like that! Usually there are many more failures than successes but a good psychic is sometimes able to detect by subtle signs when the genuine material is 'coming through' and when it is just fantasy.

GERARD CROISET'S PSYCHIC INVESTIGATIONS

Gerard Croiset, the well-known Dutch clairvoyant, died in July 1980. During his lifetime he was often able to provide valuable information which assisted the police in their investigations. Croiset was willing to be tested by scientists, in particular Professor Willem Tenhaeff, the parapsychologist at Utrecht University: as a result, several hundred cases have been examined and kept on file.

Sadly the cases often concerned the many missing children who are lost in Holland's vast waterway system. For example, in April 1963 Croiset was able to confirm that one boy, Wimje Slee, had in fact drowned near a small house with a slatted weather vane and correctly predicted that the body would be found between two bridges nearby on the following Tuesday. Croiset was loathe to work on criminal cases for fear of implicating an innocent person since, although his powers enabled him to describe the presence of someone at the scene of the crime, they did not allow him necessarily to determine whether such a person was actively involved or merely a passer-by. However, he was able to give the police clues and helped unravel several fraud cases.

When the plane carrying the young Uruguayan rugby team crashed into the Andes in 1972, parents of the boys called on Croiset to help. He was unable to pinpoint the exact area and some of his suggestions were misunderstood so there can be no claim that he assisted the eventual rescue of the survivors. However, his description of many features in the area and of the accident itself were indeed accurate. As Croiset himself warned, 'Never put too much on a psychic. I'm not a miracle worker. My psychic ability is just one amongst many others.'

Croiset regularly assisted the police — especially in their search for missing persons. In 1963 he successfully predicted just where a young boy's body would be found in the Vliet Canal. Happily he sometimes could assure families that missing relatives were alive and could even tell them when to expect them home.

SELF-TESTING

It is often suggested, and would certainly be in accordance with the model of the human mind described in chapter 1, that we are all psychic. However, it is very clear that some people are much more psychic than others and, as will be seen later, this may be the result of certain structural differences in the brain, sometimes produced by accidents to the head or other external events. I would certainly go along with the view that we are all psychic to a small extent and that we can develop this with practice. (Jahn's successful experiments did not use psychics but ordinary volunteers.) Successful business men who use 'hunches' are perhaps doing just this.

It has been clearly shown many times that subjects in parapsychological experiments are more likely to have success if they really believe that it may be possible and less likely if they consider it to be totally impossible. Gertrude Schmeidler, who did much of this work in the early stages, called such people 'sheep' and 'goats'. So, if you are to have success in your own self-tests, you should be perfectly clear that success is likely and lots of other people have also been successful. This is most important.

Also, success in psychic functioning in experiments of this kind appears to be more likely if a subject is in a happy, tranquil and relaxed frame of mind. (The tension evident in suggestion-type cases is probably the result of the fear and expectancy produced by the acceptance of the suggestions which cause the apparition, rather than the cause itself.) The conscious mind, which we might look upon as a sort of screen, should not be filled with other matters when we are going to try to detect the subtle writing or sketches put there by George (sometimes in symbolic form). George will be able to do this task more effectively if he is trained to do so — if there is regular practice together with feedback (information about success or failure). The subtle intimations reaching the conscious mind will then be more readily recognized and separable from fantasy.

In my view a way *not* to learn to function psychically is to carry out a series of card-guessing experiments, the cards being selected by using random-number tables. In the history of the paranormal this has very occasionally been successful but in my view it is just about the most effective way of extinguishing incipient psychic functioning through boredom. It is much better to try to receive impressions of interesting and attractive objects which are quite distinctive. It is also quite important to have rapid feedback — to give George information straight away after he has made the attempt to convey impressions to you.

I suggest therefore that you should proceed as follows. Sit down with a friend in a quiet familiar room. Talk about ESP and the model of the mind described in chapter 1. Remember that George probably already knows the answers to the questions you are going to put later and is perfectly capable of giving them to you if you co-operate well with him. It will probably be greatly helpful if you are relaxed and happy, a little excited that you are about to try something you have never perhaps taken seriously before and something which may be very important in your life from now on; in this way you will have an 'emotional impetus'. Then you should ask your friend to go to another room and select an object having some fairly distinctive features and a reasonable number of them, such as shape, colour, smell and texture. Your friend should put this on the table and have a good look at it. When you are ready to try and receive your impressions and your friend is also prepared you can indicate this to each other, perhaps by a shouted signal. Your friend should hold the object, turn it over, feel it, smell it (if appropriate!), and generally become well acquainted with it. At that same time you should relax and ask George to give you those impressions of it you need.

Be alert for the gradual emergence of quite fleeting impressions at first. It is most unlikely that you will suddenly know the identity of the object in one go. If you have such an overwhelming impression it is much more likely to be the result of a random guess. Imagine you are holding the object near to you and let George give you all the sensory inputs you will have when you hold it later in just that way. Ask yourself general questions about the shape, colour, weight and other features of the object, but above all, keep it general: avoid being too specific. Most important, do not make guesses as to the

ARE YOU PSYCHIC?

Think about these questions carefully; answer each one as honestly as you can.

1 Do you sometimes know when a friend is about to call or write?
a. Yes – also I sometimes phone friends and discover they are just about to phone me.
b. Only when they are already expected.
c. Often I am thinking of someone when they call but I consider this to be coincidence.

2 Do you ever experience a strong feeling or atmosphere in a place or building?
a. Yes
b. No
c. Sometimes places do have a happy or sad atmosphere but surely that's normal, isn't it?

3 Have you ever felt that you are not actually within your body any more, that you have floated away and can observe yourself from some distant point?
a. Yes, more than once.
b. No, never.
c. Perhaps, but only during dreams.

4 Have you ever had a strong premonition of some impending disaster?
a. Yes, and I was proved to be right.
b. No
c. Yes, but on no occasion did anything actually happen to justify these presentiments.

5 Have you ever been aware of the presence of someone else or heard a voice, when to all intents and purposes you were alone?
a. Yes; this sensation can be most convincing.
b. Never
c. Not really, or at least only in circumstances in which I might expect to feel nervous.

6 Have you ever been aware of objects moving mysteriously in your home or previously unknown objects appearing there?
a. Yes
b. No
c. Yes, but I am sure there is a normal explanation.

7 Do you think the level of coincidence in your life is abnormally high?
a. Yes, it does seem that way.
b. No, I do not think so.
c. Yes, but perhaps it is *being* aware of the coincidences that makes me appreciate this.

8 Have you ever seen shadowy figures or even someone (or something) apparently quite solid, which no one else could perceive at the time?
a. Yes
b. No, not to my knowledge.
c. Yes, but I am sure there is a rational explanation.

9 As a child did you have an imaginary friend who was a very real person to you?
a. Yes
b. No
c. Perhaps, but as I was so young it is a very vague memory.

10 If asked to be a fortune-teller at a fête would you expect to:
a. Have a degree of success in predicting the future of some of your clients?
b. Treat it all as a piece of play-acting?
c. Be a little nervous just in case any statement you made came true?

11 Have you ever been aware of exerting mind over matter; of making something happen by will-power?
a. Yes, quite definitely.
b. No, never.
c. Perhaps on a few occasions, but it could have been coincidence.

12 Have you ever known something is about to happen, seen it in your mind, before the event takes place?
a. Yes
b. No
c. Yes, but perhaps there were normal clues or signals of which only my subconscious was aware.

13 Do you think people have an aura? Are you aware of a spiritual element as well as the physical body when you look at someone?
a. Yes, definitely.
b. No!
c. I can't see anything tangible but I do have instinctive feelings about people, their state of health and so on.

14 Are your dreams very vivid? Do you feel you are sometimes able to direct them?
a. Yes, quite often the imagery is very real and there is an element of being able to control events, especially just before waking up.
b. I don't know because I don't remember many dreams.
c. Occasionally, especially if I have gone to bed in an anxious frame of mind. I had very vivid dreams as a child which I still remember now.

15 During metal-bending sessions on television have you had a personal experience of the phenomena, such as a spoon bending or your watch stopping?
a. Yes
b. No
c. No, although I did experiment at the time.

16 Does electrical equipment in your household wear out or go wrong abnormally quickly or often?
a. Yes
b. No, not that I have noticed.
c. I am unsure but I will keep a check on this in future.

17 Have you ever heard voices in your head telling you what to do?
a. Yes, and they are quite distinct from the normal thought processes.
b. No, and if such a thing happened I should keep quiet about it!
c. Only when my conscience tells me something is right or wrong or my 'other self' intervenes.

18 When asked to write an imaginative story have you ever felt:
a. That someone else has taken over the pen and written it for you so that you do not even know what you are writing until afterwards?
b. That it is impossible to start; your imagination will not work?
c. That sometimes it is as though the story writes itself and your ideas seem to come from thin air.

How to score: A high score of a *answers means you are psychic and must surely already be aware of the fact. A majority of* b *scores means that you are not psychic and are determined to remain that way. Mainly* c *answers suggests you may well have psychic tendencies but have not yet allowed yourself to recognize this.*

identity of the object. If you have any fleeting impressions of shape sketch them on a pad; also write down notes of your other impressions.

At the end of, say, fifteen minutes put aside your pencil and ask your friend to bring in the chosen object and place it in your hands. Compare your impressions as written and sketched with the characteristics of the object.

If you carry out the experiment in the way just described you will have the best possible chance of early success. You will have working for you — telepathy (co-operation of your friend's George with yours), clairvoyance (George's own powers) and precognition (George's knowledge of the sensory inputs when you hold the object in your hands later on).

A careful and thoughtful chat with your friend at the end of the experiment will be very helpful in recognizing the true intimations of the object in future and ignoring the wild guesses and extraneous impressions which are perhaps the results of sensory inputs from the ambience and false trains of thought. Above all remember that George is not logical; he is a-logical. Do not use your reasoning powers in any way: do not try to guess what your friend might select: just let George float the impressions up to you and observe them without analysing or deducing.

Try a distant-viewing experiment. Draw your impressions of what your friend is seeing and write down any general comments.

TEST YOUR ESP WITH A FRIEND

1 Before your friend leaves the room sit quietly and discuss your ideas on ESP

2 Your friend should go into another room to select and think about a suitable object.

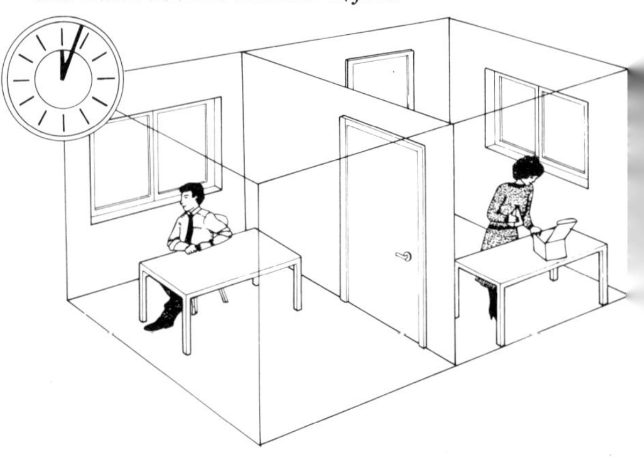

A good way of practising ESP is while looking for somewhere to park the car. Why not let George try to help you next time you need a parking place by driving in the direction of his intimations instead of using your reason and logic? You might find that you have on the whole much more success than you expect. Your sceptical friends will of course say it is all coincidence: you must make up your own mind about such things! If they try to damage your faith in George, much better not to talk about your experiments with them. Do everything you can to help George and to train George to work with the conscious part of you — to the general improvement of your life.

A final thought! If your most sympathetic friend is going to another part of the country and the opportunity presents itself to you both, why not conduct an experiment on distant viewing? If the timing were appropriate you could arrange to be sitting in a quiet familiar place ready to receive impressions at the time when your friend was walking around a site which you had never seen. All the remarks made above in regard to your strategy would apply except that you could well have a much wider range of impressions involving wind, cold, water, trees and so on.

Whatever experiments you choose to try first, remember to be positive and optimistic; others have succeeded. Good luck!

The 'agent' should let all his senses absorb the surrounding details while the 'percipient' (shown on the left) relaxes and hopes to receive them.

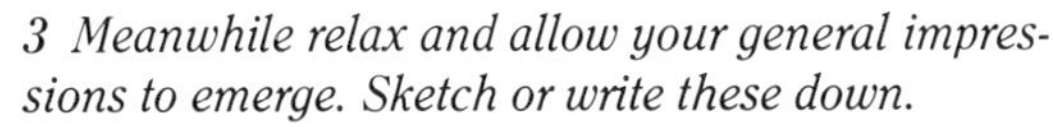

3 Meanwhile relax and allow your general impressions to emerge. Sketch or write these down.

4 Ask a friend to return. Now compare your notes and sketches with the actual chosen object.

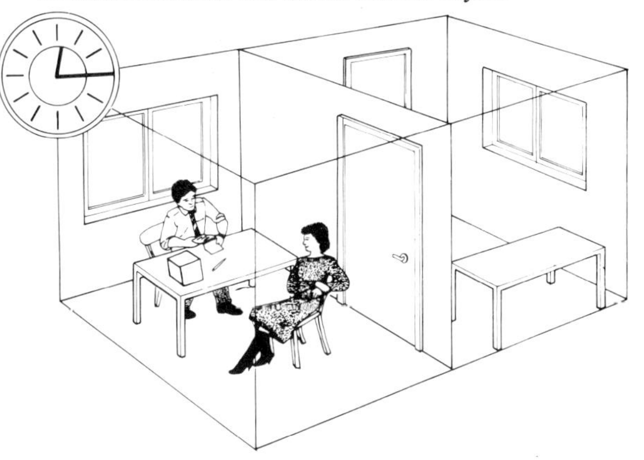

4

THE SÉANCE

Most people's ideas of séances and mediums are derived from films and plays such as *Blithe Spirit* and always include a medium like Madame Arkarty. The picture is of a group of tense sitters gathered round a table with their fingers resting on the surface and touching in a circle; there is a dim light and the medium starts the proceedings by saying, 'Is there anybody there?' I have lost count of the number of different séances I have attended and I never remember one quite like that!

Séances are held for different types of phenomena — mental and physical; and not all psychics are Spiritualists, though most are. Psychics, sometimes called 'sensitives', are individuals whose minds, in our model of an iceberg (see page 11), would have a crack allowing normally unconscious material (material below the sea line) to become overt in the form of hallucinatory impressions or pseudo-sensory inputs: in other words the psychic experiences hallucinatory visions and voices, or sometimes feelings. George who, in the model, is pushing the material up into the conscious mind, often has to use symbols which must be interpreted by the psychic percipient.

A visit to a medium is more likely to look like this!

How the uninitiated might imagine a séance.

The Spiritualist interpretation of this phenomenon involves discarnate people (people in the 'next world') putting the material before the psychic. The psychic is called by Spiritualists a medium because he or she is assumed to act as a medium between this world and the next. According to the Spiritualist theory, when mediums' minds and bodies are used in this way, they are called mental mediums. In other words a mental medium is a person receiving data ostensibly communicated from the dead via the mind and body of the medium.

There is a different kind of medium — and a different kind of phenomenon — from this. When the production of ostensibly paranormal *physical* effects takes place during a séance then the medium is called by Spiritualists a physical medium and the phenomena, which are assumed to be produced by entities in the 'next world', are referred to as physical phenomena.

Let us consider these two types of séance — and start with an example of mental phenomena.

VISIT TO A MENTAL MEDIUM

Just before Christmas 1970 a friend of mine who lived opposite died of lung cancer. He was a tall thin man who worked as an accountant. Most mornings he drove me to the station on his way to work and we often discussed parapsychological matters.

In the spring of 1971 his widow came across the road to see us and during conversation said this to me. 'You and my husband often used to discuss psychical research and, amongst other matters, the possibility of human survival of bodily death. I do not know whether my husband survives or not, I think perhaps he does. But if he does perhaps we should give him a chance to prove it. Will you please arrange a sitting for me with a really good medium?'

This was a splendid opportunity for a completely controlled experiment and I immediately complied. I telephoned a well-known medium whom I knew well (Ena Twigg) and asked her to arrange an appointment for a 'someone I knew'. (That is the only information I gave her.) She gave me a date and time and I passed these on to the widow across the road. Before the sitting I gave my friend's wife careful advice on what to say and how to respond during the sitting. She was to be warm and encouraging but to give absolutely nothing away. She did not have a tape recorder but the elder of her two daughters was a skilled secretary so she accompanied her mother and took along with her a pad and pencil. The two of them went along to keep the appointment and I asked the daughter to record every word spoken by herself, her mother and the medium. The following evening the daughter brought round to me a carefully typed manuscript containing an account of the séance.

The medium described a tall thin gentleman whom she said they had brought in with them: she could see him and they could not. She stated clearly that he was the husband of the elder sitter and that he himself was giving her this information. She could see him with her clairvoyance and hear him with her clairaudience. All this occurred in the medium's pleasant sunny room in July 1971.

As is usual, the medium spoke many times as though it were the deceased husband speaking. At other times she spoke as from herself. 'This is someone who has never communicated before.'

'*...I feel so ill. I feel so ill. I can hardly speak, I feel so weak. I can't breathe very well. I have not been over here very long. Less than a year ....*'

'I can see a big letter D and then two big capital D's.' These are the initials of his Christian name and the pet name of his elder daughter. This makes it quite clear that the picture of a deceased communicator speaking to the medium is not correct — if it were so he would hardly need to produce letters in this way.

'*...Didn't want to die in hospital. They let me come home, but I had to come back. They couldn't cope and I had to be propped up because I could not breathe. They gave me oxygen and an injection....*' All this was perfectly accurate.

'I see him riding a horse' was not understood by the sitters, but the other statements made were correct. 'You have a Persian carpet somewhere in your house. Take care of it. He is particularly fond of it.' That last was obviously from the medium but it was quite correct. There were many statements about their life together, their visits to the country, their seeking for old things to buy for their house, his love of paintings, his fondness of his mother, about the women in his family. 'There must be another child.' '*I want to send my love to A.*' There was another child (not present) and her initial was A.

'*...And my swallowing was so difficult. My mouth was so dry and they just moisten it. They used to put ice on my mouth.*'

'Married twenty-three years.' This was correct but it was twenty-two when he died. 'He mentions a photograph of his father with a moustache.'

'*We resemble each other.*' This was correct too. '*There is a dinner suit and three evening shirts. See if someone can use them. Don't leave them there.*' This was all correct and was typical of him. '*See to the apple trees. They were good trees. We have got a lovely garden.*' All this was correct. '*...Somebody has a walnut tree.*'

This latter was not understood at the time of the sitting but when the family showed the manuscript to some friends nearby they understood it well and said that the walnut tree was actually the one in their garden and had been discussed with him before he died, but the rest of the family were not there at the time.

'It would be stupid if all that remained of a man of my age was a handful of dust and a memory. Given the opportunity, I could be more vital and forthcoming in this world than I could be even when I was on earth. So continue: think we should do some kind of experiment. If you will co-operate, I will. I do like to feel you are involved and directly concerned. So we will pursue it.'

This gives the general flavour of it all. Some of the statements made were very evidential but others were rather vague. When I received this manuscript I listed all the statements made in three columns: correct, incorrect and not understood. The medium made 172 separate statements in a sitting lasting one hour and of all the statements eighty-five per cent were correct. If we omit the statements which were not understood or vague and merely count the correct and incorrect, then ninety-five per cent of them were correct. It is important to realize that, with the exception of the reference to a walnut tree, all the correct statements were recognized as such by the sitters at the time.

Let us now discuss the various ways to look at this experience – the models which represent what was happening. First there is the medium's explanation of her experiences which we might have some difficulty accepting exactly as she gives it. We shall then look at facts which support the iceberg model of the mind (including George) which was offered in chapter 1 (see page 11) and apply this to the facts of that séance. It turns out that things are not nearly so simple as they appeared at first sight.

The medium would say something like this: 'You may think you are just that physical body I can see, but you are not really. You have other bodies, in particular an astral body. I can see it with my clairvoyance and it is made of subtle matter as yet unknown to science and it projects all round. The colour of the projecting part, the 'aura', depends on

Sant Bernhartz predige von der menschlichen hartselikeyt.

Were saints' haloes part of their aura?

your emotions. If you are a very spiritual person then the aura round your head tends to be purplish which is the reason for the halos round the heads in medieval paintings of saints. When you die your physical body disintegrates but you are unchanged in your astral body and you go into the astral world – which is also made of subtle matter unknown to science and interpenetrates this one – and there you meet your deceased relatives and friends. Normal decent people go into the Summerland, which is very like this earth only much more beautiful, and selfish unpleasant people go into a very gloomy place which has led to the stories of hell. When we have a séance and if we are lucky, deceased relatives and friends of the sitter will come to my room and I see them in their astral bodies with my clairvoyance and hear them with my clairaudience. I have no way of compelling them to come.'

So what is wrong with this explanation? Before going into that, let me say first that mediums are amongst the most honest and genuine people I know. One very rarely meets a fraudulent medium. However, they are not scientists and are not always aware of the shortcomings of their rather literal interpretation of the experiences that they do genuinely have.

HYPNOSIS

First a few more words about the dramatizing capabilities of the unconscious mind (George). In 1959 there was a series of programmes on UK television about hypnosis. They were presented by a well-known London psychiatrist; the hypnotist worked as a doctor in one of the other London hospitals. The hypnotist had brought along two good deep-trance subjects with him who were nurses in his hospital. The programme took place in a large BBC studio with a few chairs near the camera at one end for the audience. At the other end in the corner there were several rows of empty chairs near a closed door. The presenter asked the hypnotist to put one of his subjects into deep trance, which took only a moment. He then asked him to give her a post-hypnotic suggestion (which she would forget) that people would come into the studio and sit in the chairs at the back. Having done that he woke up the subject and brought her completely back to normal. She had no memory whatever of the suggestion that had been made. The presenter then asked her if she could see anything unusual in the studio. 'Yes', she said, 'I see people coming into the studio in the middle of the programme and sitting in chairs at the back.' The presenter then said to her, 'Will you please go over there and ask one of those people to come forward to answer a few questions?' She then walked to the back of the studio, had a somewhat one-sided conversation with an empty chair and came forward to the front again. She said 'I have brought 'Mr Brown'. [I have forgotten the name used.] He is quite happy to answer a few questions.' It is important to realize that 'Mr Brown' looked perfectly normal and solid to her and she had no reason to doubt that he was not just a new member of the audience. The presenter then asked her to inquire of Mr Brown what he was doing in the studio and how he had travelled there. She did this and Mr Brown replied, she reported, 'I am here because I am interested in hypnosis and I drove down the M1 motorway this morning.'

At this time a BBC technician brought in a chair and placed it in front of the audience. He was holding a Polaroid camera. The presenter then asked the subject to invite Mr Brown to sit in that chair, which she did, and the technician took a photograph of the chair. Pulling out the film and peeling off the backing he handed the photograph to the presenter who showed it both to the television camera and to the subject saying 'What do you see?' The subject said, a little puzzled, 'Why, it is Mr Brown sitting in that chair, of course.' It merely showed an empty chair. The BBC had earlier placed in front of the audience a platform weighing machine. The presenter then asked the subject to tell Mr Brown to step on to the platform of the weighing machine and asked her to state his weight, which she did, having to lean round the entirely non-existent Mr Brown in order to read the scale.

At the end of the programme the presenter explained that we all have dramatizing machinery like that in the unconscious mind; this machinery is perfectly capable of filling a large hall with fictitious people, each with a name and address, a description and a life history. There is no doubt about it. This machinery (George) produces our dreams and may be receptive to the suggestions of a hypnotist and able to carry out instructions received under hypnosis, within certain limits. The question that will naturally occur now is, of course, 'So who made the suggestions in that séance with the accountant's widow and how is it that so many of the statements made were correct?'

COMMUNICATION

First let us have another example to illustrate what went on at another séance. Dr S.G. Soal (a former President of the SPR) was at a sitting some years ago with a medium called Mrs Blanche Cooper. At Mrs Cooper's séances a voice, not her own, appeared to originate from mid-air. (For information on a possible mechanism see page 49 . Here the concern is with the factual material received in this way.) A psychologist would probably suggest that the source of the information was a secondary personality of the medium, just as if the voice were emanating from the lips of the medium in trance. Most secondary personalities state that they once lived on this earth and are now in the next world, their work being to pass on information from deceased communicators. (Sometimes such communicators claim to be speaking directly — in which case the voice usually alters.) However, not many control personalities — as they are usually called — can be traced from facts available here. In Dr Soal's case the control personality and Dr Soal's deceased brother Frank both appeared to be passing on what looked like very good evidence concerning his brother — facts which Soal had long forgotten or, in one example, had never known.

During this series of séances another communicator appreared and stated that his name was John Ferguson, originally from Brentwood in Essex, who sent love to his brother Jim and his wife. He gave his age, the date and manner of his death and the general subject of his work. Soal later remembered a James Ferguson, another boy at school with him in Southend, and vaguely recollected facts he had given about his father. Soal had lost sight of James Ferguson years before.

Soal now visited Brentwood (for the first time) to acquire facts about the locality. He formed in his mind a series of 'fruitful conjectures' about John Ferguson. He received back from the voices at the next séance some of the facts he had noticed or speculated about. Soal again visited Brentwood and interviewed the postmaster. It was clear that no persons having the names given in the séances had ever lived in the roads also named. Soal invented more theories around the personality of the fictitious John Ferguson. Interesting and ingenious allusions to these appeared at the séances. Later on John Ferguson said that he had lived in Glasgow, where he had received further education and, eventually, been buried. Soal looked at a map of Glasgow and

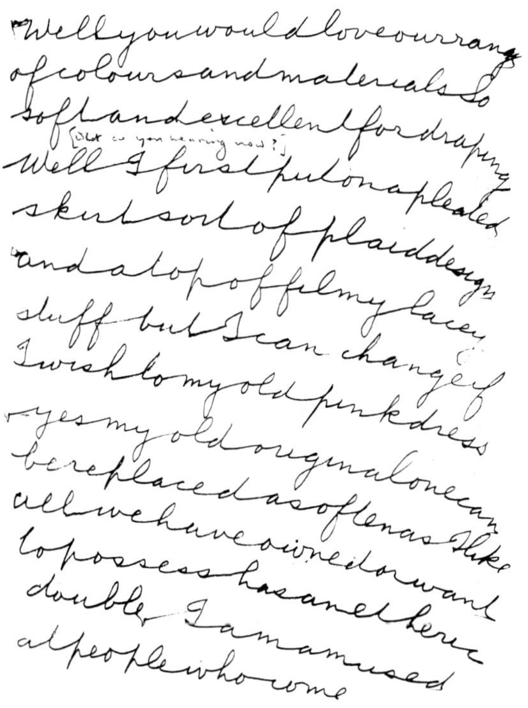

An example (and transcript) of automatic handwriting with the words flowing into each other.

...well you would love our range of colours and materials, so soft and excellent for draping...

'What are you wearing now?'

...well I first put on a pleated skirt sort of plaid design and a top of filmy lacey stuff but I can change if I wish to my old pink dress. Yes, my old original one can be replaced as often as I like. All we have owned or want to possess has another double. I am amused at people who come...

then asked for street names in Glasgow and the whereabouts of the cemetery where John Ferguson was buried. He received all this information from the séances as requested and then wrote to the keeper of the cemetery who confirmed Soul's suspicion that no people with the names stated were buried there around those dates.

Summarizing, it seems the new communicator, John Ferguson, gave many facts to prove that he really had lived on earth, yet the only life he had ever lived had been in Dr Soal's imagination. One is perhaps reminded of Charles Dickens who said that sometimes the characters in his books were so positive that they took over the writing from him.

The Irish poet W.B. Yeats was once having a sitting with the automatic-writing medium Geraldine Cummins. Her 'George' used her muscular system to produce writing in order to make the material overt. In this instance the writing was describing an old castle and the people who lived in it. She read it with interest and said to Yeats, 'Shall I let this go on? Are you interested?' Yeats replied, 'I certainly am: .that is the plot of my new book.'

There is no doubt that information in the mind of a sitter somehow penetrates into the consciousness of a psychic and is dramatized in various ways by that unconscious machinery we have, for convenience, called George.

There is another way that George sometimes uses to make material overt and that is the Ouija board. Here the recipient of the 'communication' puts the fingers gently on a pointed piece of wood called a traveller and the pointer slides over the polished board pointing to the various letters of the alphabet which are inscribed upon it, together with the words Yes and No. Another version of the Ouija board is the well-known party game in which a polished table is used with an upturned glass. The letters of the alphabet are again put around the edge and the sitters place the tips of their fingers gently

Matthew Manning 'received' some dozen or so drawings which turned out to be reproductions of Albrecht Dürer's work or unknown drawings in his style.

on the bottom of the upturned glass (see also page 103). If success is achieved the glass moves to the various letters, spelling out messages, while all the sitters stoutly maintain that they are not pushing it. (Of course they usually are, but unconsciously.)

Some mediums are able to go into trance and produce what is called automatic painting or drawing and these are sometimes very like the artistic productions of famous deceased artists, the medium explaining that this is a way the deceased artist is using to prove his or her survival.

ASSESSING THE SÉANCE

In the light of all these facts let us now assess the material of the séance held with the widow of the accountant and her daughter.

The first thing to say is that it is simplistic to suggest that the deceased husband came in his astral body and was observed by and talked to the medium who passed on her experiences. That clearly will not do in view of Dr Soal's semi-imaginary communicator and Geraldine Cummins' passing on of Yeats' plot. The true picture is clearly more complex. In the model we have of the mind (acceptable to most if not all parapsychologists) we imagine George acquiring the information in some way and floating it up into the conscious mind of the medium as hallucinatory visions and voices. So what is the likeliest source of all this material? Surely it is the memory stores of the sitters.

In the case of a very successful séance such as this one, there appears to be some sort of a 'fusion' of minds at the unconscious level and the medium (quite unknown to herself) is presented with the information in the sitter's memory store. George, in the way now familiar to us, dramatizes all this material so that it appears to come in the way the medium describes. Just about all the information we have in our normal life has been acquired through one or more of the five senses and there is no mystery about why George selects imitations of the operations of those senses to make the information overt.

In point of fact, scientific evidence that telepathy sometimes occurs is exceedingly good (there is no doubt in my mind that this is so) and the likeliest explanation of the many successes of that particular séance is telepathy — the presentation to the medium's George of relevant items from the sitters' memory store. George, in collaboration with the sitter's George, perhaps selects the relevant material before dramatizing it.

So the material acquired at such a séance can be very confusing and requires a great deal of experience to understand. There are clearly very good examples which are probably the results of telepathy from sitter to medium. Some of the dramatiz-

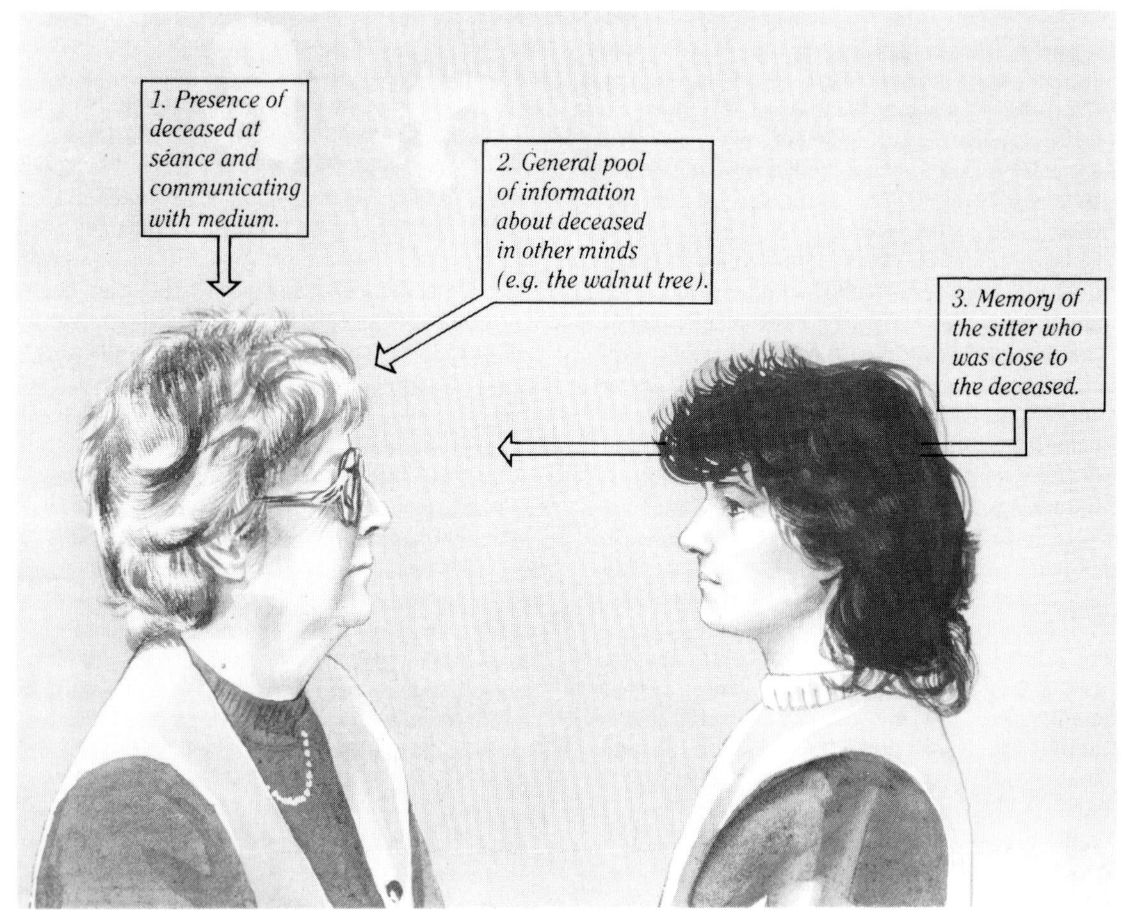

The medium may unconsciously be drawing on several different sources of information.

ations will undoubtedly be based on the medium's conscious and unconscious guesses as to the nature of possible communicators and perhaps, if the séance is particularly good, there may be signs of a genuine communicator somewhere in the background. May I interject that the scientific evidence for survival is, in my view, exceedingly good.

The perceptive reader will remember the item communicated about the walnut tree which was not in the memory store of either sitter and was marked 'not understood'. The family friends nearby pointed out that the tree was in their garden and had been discussed with the deceased husband before he died, the sitters not being there at the time. So what is the explanation of this?

There are considered to be two and only two explanations of the walnut-tree episode. The two hypotheses are, respectively, 'survival and communication' and the 'super-ESP' hypothesis. In the super-ESP hypothesis we have to imagine that George (the medium's George) is able to discover facts from the mind of anyone in the world, whether she has ever met them or not, and these must be facts which are relevant to the deceased husband. George must then be imagined to dramatize these facts as though they are coming from the deceased husband — all especially in order to deceive us about survival. I know of no evidence of any kind that ESP is so all encompassing and wide ranging. However, I do have evidence (see below) which appears to indicate most persuasively that human beings do survive bodily death — and the alternative explanation of the success of this séance, especially the walnut-tree episode, is that the deceased husband was in some manner, which is not in any way understood, present and communicating.

It is very clear that all these factors and possibilities must be borne in mind when considering the results of a séance. Clearly a great deal of experience is very helpful when doing this.

A final remark: Sometimes mediums will give facts, as from the discarnate communicator, which lie in the future and occasionally the events described do actually happen. It is clear that, as mentioned in chapter 1, the model of the mind involves space and time seeming to behave rather differently as the consciousness moves downwards in the iceberg and out into the sea. In fact, once in the sea, space and time appear to be merged into an 'eternal now'. So George is able occasionally (it does not happen very often) to acquire material which, so far as ordinary physical-world consciousness is concerned lies in the future. This can be dramatized to appear as if from a discarnate communicator.

Not all psychics are Spiritualists, as mentioned earlier, and some of them merely use their ESP to attempt to acquire impressions of the future. They are occasionally successful but usually there is no way of telling until their predictions come about. Perhaps some of the prognostications are dramatized from the wishes and plans of the sitter. There is no doubt whatever in my mind that obtaining information by psychic faculties is just about the most unreliable way of obtaining such information! One never knows whether it is correct until it has been checked. People — and there are many such — who organize their lives on the basis of what psychics tell them are perhaps behaving somewhat foolishly.

EVIDENCE FOR SURVIVAL OF BODILY DEATH — THE CROSS-CORRESPONDENCES

It should be clear by now that much of the paranormal material I have described which appears to come from a discarnate communicator probably does not do so: it is dramatized by George to appear that way. However, there is what seems to me some very much better evidence for survival and communication and this is provided by the 'cross correspondences'. We must briefly consider this.

Imagine that you are an informed psychical researcher who has died; imagine further that you find yourself in the next world. You discover what it is you need to do to communicate with your friends and relatives through a medium. What would you communicate to indicate in the best possible way your continuing existence?

Clearly, it is no use merely sending details of your life on earth or of your emotions and attitudes to your relatives because anything that they immediately recognize as being true could well be accounted for by telepathy from sitter to medium, the material being dramatized by George to appear as if from you. No; that will not do; you must think of something better.

Myers, Gurney and Sidgwick, friends at Cambridge and some of the founders of the SPR, appear to have found themselves in just that position and they devised what we now call the 'cross correspondences'. Their scheme was as follows. They selected stories from the classical Greek literature, in which Myers especially was an expert, and constructed puzzles based upon them. They communicated one piece of a puzzle through one psychic and another piece through another psychic in a different part of the world. In the best cases the psychics did not know each other and had a negligible knowledge of classical Greek literature. In addition each psychic did not understand the meaning of the fragment she received. To provide clues to the association of the two fragments, a third psychic in yet another part of the world provided allusions which, in an oblique way, pointed to the connection between the first two pieces. Each piece was completed by a message, such as 'This is F.W.H. Myers; please send this to The Society for Psychical Research in London.'

When all the pieces were received they were found to form a consistent whole which was understandable (to someone familiar with Greek classical literature) and characteristic of the communicators. Often they showed distinct signs of the personal idiosyncracies of the communicators.

The ladies who produced the earlier material were not professional mediums but had discovered that they could produce automatic writing. They included Dame Edith Lyttelton, Mrs Holland (a pseudonym for Mrs Fleming, the sister of Rudyard Kipling) in India, Mrs Willett (a pseudonym for Mrs Coombe-Tennant), Mrs Verrall (lecturer in Classics in Cambridge University and wife of one of the later communicators Dr A.W. Verrall, a distinguished classical scholar) and her daughter Helen Verrall. After the first signs of the correspondences between the scripts care was taken that the writers did not know the contents of the scripts of the others but all were sent to the SPR Research Officer.

In later work, a professional medium being studied by the American SPR, Mrs Piper, was used by the ostensible communicators to supplement the material from the others.

Material of this kind was produced in large volumes for some thirty years and provides by far the most persuasive evidence for survival. The communicators explained that they had devised the scheme in order to eliminate telepathy between the automatic writers; this they did by expressing the ideas in so veiled a form that each writer wrote her own share without understanding it. In addition they used material few of the writers knew anything about. Also as each writer produced only a piece which was incomplete and about material of which she had no normal knowledge, it was not understood by her.

There appear to be two, and only two, explanations for the cross correspondences. One of them is that the surviving communicators devised the scheme (it was certainly not devised on earth). The other is the so-called super ESP hypothesis (briefly mentioned earlier on page 47). The super ESP hypothesis involves the whole scheme being devised at the unconscious level of the mind or minds of the various psychics, who would in some way have to discover all the material which was necessary from the classical Greek literature of which they had no prior knowledge — except for Mrs Verrall and the puzzles continued to appear after she had died. Having devised the puzzles at the unconscious level then they had to be put through in the way that they appeared entirely in order to deceive us. There is, so far as I know, no evidence whatever that ESP is so wide-ranging and comprehensive as this. I choose the survival hypothesis.

There is one further hypothesis which should be mentioned — if only to be dismissed: that is that all the automatic writers and the distinguished scholars who sorted out the puzzles carried on a great deception of the public for some thirty years.

Let us complete this piece with some sentences Myers appears to have communicated through Mrs Holland in 1904: 'If it were possible for the soul to die back into earth life again I should die from sheer yearning to reach you to tell you that all that we imagined is not half wonderful enough for the truth.' Through Mrs Piper he appears to have communicated, 'I am trying with all the forces ... together to prove that I am Myers....' and again through Mrs Holland, 'Oh, I am feeble with eagerness — how best I can be identified'.

PHYSICAL PHENOMENA

Physical phenomena are paranormal occurrences which everyone can see and hear. Let us start with an example from my own experience.

The cardboard cone floated up and circled my linked hands several times.

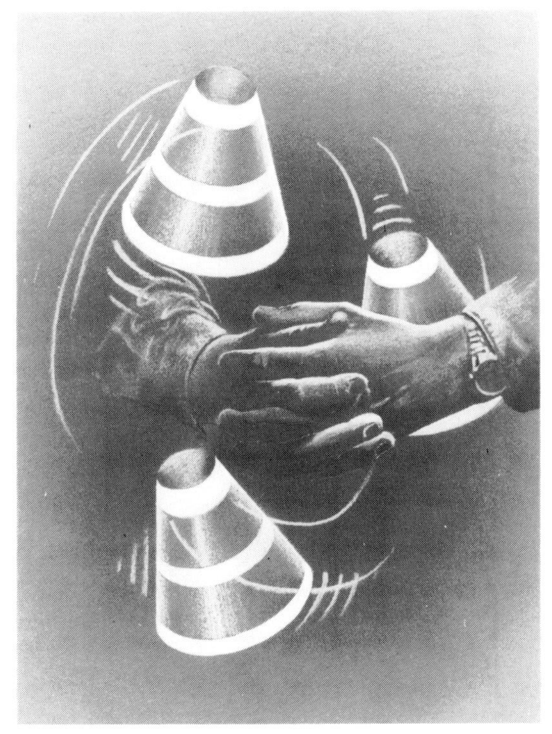

Some years ago I attended a séance in north London. There was a circle of Spiritualists with a trance medium and they had kindly permitted me to attend. At the beginning of the séance the medium went into trance. The light was still on; it was a red light as most Spiritualists seem to think that the 'ectoplasm' is damaged by white light. The result was that I could see clearly the circle of Spiritualists – with the medium a part of this circle. On the carpet in the middle of the circle, with everyone well clear of it, was a cardboard cone painted with luminous paint; this the Spiritualists called a trumpet. (It is claimed that the purpose of the trumpet is to enable the discarnate communicators to build an 'ectoplasmic larynx' within it, through which they are then able to speak.) The medium – or rather the control personality – asked me to link my fingers together and hold my arms out in front of me. This I did. The cardboard cone then floated up into the air and encircled my linked hands five or six times before floating back on to the carpet again. The control explained that this was to show me that the trumpet was not attached to any invisible rods or strings. Remember that I could see what was going on quite clearly and everyone was sitting quietly in their chairs.

Thereafter the lights were turned out and I was unable to observe what was happening. However, it was quite amusing when the trumpet (visible because of the luminous paint around it) appeared to float in front of me and a voice apparently emanated from it stating, 'Allo, mate. I'm dead but I won't lie down!'

It is important to understand what is considered to occur during a séance for physical mediumship including various types of phenomena. It will then be possible to look critically at what is happening and to consider some of the scientific work that has been done on this type of phenomena. I shall also give some of my own experiences.

List of Possible Physical Phenomena

1. Cold breezes can sometimes be felt.

2. Whispered voices emanate from the different parts of the room.

3. Ectoplasm may exude from the bodily orifices of the medium. This can form into disembodied limbs and faces – or figures (sometimes exotic characters) which may be able to talk. Ectoplasm can be seen by the light of luminous plaques or in infra-red light.

4. Little sparks or faint lights appear in the air.

5. The investigators may feel the touch of invisible hands.

6. Objects not previously in the room, (apports), may appear – sometimes on the laps of the sitters. Often these are flowers.

7. There may be table rapping and levitation. The latter may be accompanied by a weight change in the medium and/or sitters.

8. Musical instruments may be played and the curtains billow.

What happens at a séance for Physical Phenomena

The medium goes into trance and gives, as from the 'discarnate control', some sort of sermon consisting usually of a series of platitudes concerning helping one another and how we are being helped from the next world. There is normally nothing evidential in such sermons which are, it seems to me, not of much value. After this the interesting things begin to happen — if the sitters are lucky. Usually they are enjoined by the control to sing and they do so at the tops of their voices. (Unfortunately not all Spiritualists attending séances of this kind have a musical ear and the result may not always be pleasing to a musician!) The purpose of the noise is to 'develop power', whatever that may mean, and if everything goes according to plan the first phenomena to be noticed will probably be cold breezes. Sometimes little sparks of light can be seen in the air and occasionally whispered voices will emanate from various parts of the room. At this point objects may sometimes be found on the laps of sitters; these objects, it is claimed, were not previously in the room, the doors and windows of which are fastened. They are called 'apports' (from the French *apporter* which means to bring) as they are brought from

Ectoplasm may form into faces which can be observed in the light created by luminous plaques.

elsewhere into the séance room. Sometimes apports are flowers and a séance room can be filled with a mass of flowers at the end of the séance.

The *pièce de résistance* is materialization. Figures are supposedly formed from rather tenuous greyish substance called by a distinguished French scientist (Richet) ectoplasm; this is exuded from various bodily orifices of the medium. This ectoplasm forms itself into faces and figures which are able to talk and be observed by sitters, the observations being made in the light created by luminous plaques. (These are simply square pieces of plywood painted with luminous paint, each carrying on its rear side a handle.) Sometimes the faces and figures are observed in artificial light, usually red. The materialized figures will pick up the plaques from the floor and then, holding them close to their faces, become visible to sitters.

At the end of the séance the medium will, often with much grunting and gasping, come out of trance, the lights will be switched on and he or she will be helped from the room exhausted.

What are we to make of all this?

Many people have a rather oversimplified mental picture of what a materialization séance is. The picture they have is of discarnate people coming to the séance room in their astral bodies and clothing themselves with ectoplasm collected from the medium's body. They then imagine that the discarnate people are firmly in the physical world and can see and speak. As many people, particularly the bereaved, would very much like to see and speak to their deceased nearest and dearest they are prepared to pay a fair sum to attend such a séance. The result is that a medium who is not honest has a very strong motive for faking the whole thing. Many such séances have been faked in the past and many probably are being faked today. The extreme reluctance shown by some mediums to permit any investigation by a parapsychologist who is well experienced in the subject tells its own story.

Two early SPR investigators, Myers and Davey, did experiments to see how crude a fake séance could be while still deceiving the sitters and ran a number of séances which were faked from beginning to end but without telling the sitters. They wrote a paper giving their results which has become a classic. It is quite astonishing how a hoarse whisper combined with a bit of cloth waved in the air and vaguely seen will be recognized as a deceased relative. If the sitters happen to be bereaved and with great readiness to believe, they are perhaps even more easily deceived. If they are Spiritualists and take a semi-religious approach to the whole business then they are especially open to deception as it would be considered almost sacrilege to doubt.

INVESTIGATION BY SCIENTISTS

A great deal of good and soundly based scientific work has been carried out to study the physical phenomena of séances. Some of the best ever was done in Paris near the beginning of the present century. It was organized by the professor of psychology at the Sorbonne (University of Paris) and the researchers included Henri Bergson, Pierre and Marie Curie and Richet (Nobel laureate in physiology), all professors at the Sorbonne. They studied a famous Italian medium called Eusapia Palladino. The investigators had a most complicated system of instruments designed to monitor phenomena and record any physical occurrences. Also the monitoring was in a separate laboratory thereby removing the investigators from any influences present in the séance room. The medium's chair was on a weighing machine and tables were electrically wired so that their movements could be registered. Every noise and comment in the séance room was recorded and a reasonable light was maintained, one that was variable as desired.

During these tests a large number of physical phenomena were observed including tables levitating and raps produced some distance from the medium – whose weight increased by the weight of the table when it was levitated (showing that the force was coming from her). There were other phenomena like curtains billowing out, a musical instrument being played, investigators touched and faint lights seen. Sometimes hands and arms were observed which did not belong to anyone present.

Sadly Pierre Curie was killed in a traffic accident before these experiments were completed. The investigators gave the results of their work and the famous astronomer Flammarion suggested that levitation should be in no further doubt. It really is difficult to see how better tests than these could have been arranged. Nonetheless, the SPR refused to endorse them and sent their own team of three experts in the detection of fake physical phenomena and conjuring to Naples (where Eusapia Palladino lived). They took a room and equipped it with their own materials, buying objects which were required such as musical instruments. Eusapia brought her own 'cabinet' — in the form of a curtain across the corner of the room — behind which the objects could be placed on a small stand and she brought her own table which they thoroughly examined.

From their report it is very obvious that those three investigators were hostile and doubting in the extreme. They certainly did not wish to believe; quite the contrary. Though they observed all the phenomena they felt that they must have been subject to hallucinations or had not observed clearly. They questioned their senses rather than their experiences. Doubt remained! But I wonder whether there was good reason for it. At this distance it looks more like prejudice.

Miss Kathleen Goligher, a medium whose psychic powers were studied by Dr Crawford.

TABLE-RAPPING SECRETS.

LEVITATION THEORY.

SCIENTIST'S FATE.

Mr. Wm. Jackson Crawford, D.Sc., lecturer in mechanical engineering at Belfast Technical Institute, on whom a coroner's jury at North Down yesterday returned a verdict that death was due to poisoning—he had been found dead on the rocks at Bangor, County Down—had achieved some publicity by a series of psychic experiments with a girl medium, the results being embodied in two books which have had a large circulation among spiritualists.

These investigations were designed to show that table-rappings and other spiritualistic manifestations are produced by "matter" extracted from the body of the medium, and manipulated by an unseen agency, and that the "matter," after the manifestations, returns to the body of the medium.

Dr. Crawford's medium was Miss Goligher, who has never acted as a professional medium. It was claimed that the experiments were strictly scientific, and were carried out in conditions which precluded the possibility of fraud.

During some hundreds of séances, in which Miss Goligher sat in a room, six other persons being present and clasping hands, in a fair red light, some matter, of undefined composition, was taken from her body and utilised by some unknown agents to levitate [to raise without visible physical means] a table 4 feet or more into the air.

REDUCING WEIGHT AT WILL.

The girl was placed on a weighing-machine, and when the table was levitated its weight was registered as though it were actually on the scales. Dr. Crawford found that while the girl lost weight during a séance, when it was over and the "matter" returned to her body her weight was restored.

He asserted that once when he asked the "operators" to extract as much weight as they could from Miss Goligher's body, her weight was reduced by 20lbs. For a few seconds, he declared, she lost 54lbs., about half her normal weight.

As to the identity of the "operators" Dr. Crawford said: "Nearly every real investigator has now come to the conclusion that these entities are what they claim to be, the spirits of human beings who have passed into the next stage of existence."

Dr W.J. Crawford's scientific investigations of séances aroused public interest.

Another investigation was conducted by an academic staff member of the Queen's University of Belfast, a certain Dr W.J. Crawford, who was a mechanical engineer. He did some useful work with the medium Kate Goligher. She had a home circle and Crawford was very impressed by the probity of the members of her family who were also the members of the circle. The séances were held in a reasonable light, there was no possible reason for deceit as the séances were being held privately, and Crawford managed to win the confidence of the family who allowed him to attend their séances. He was permitted to do just about any tests he liked while the séance proceeded.

Crawford, in those earlier days of the First World War, used a phonograph to record sounds and he found that, although this instrument was placed a distance away from the medium, the control produced loud noises which caused the same result as a voice very close to or in the horn.

Most of Crawford's work involved using the principles of mechanical engineering to try to elucidate the mechanisms of the table levitation. He arranged for the medium and the sitters to be seated on weighing machines and discovered that when the table was levitated the medium gained in weight by almost the weight of the table, some of the weight being distributed amongst the other sitters. 'Ectoplasm' appeared to be forming rods emanating from the medium and bending upwards underneath the table, so lifting it. He found that he could actually put his hand through the space where the rod must have been without actually feeling anything or stopping the levitation. When he photographed the circle in infra-red light he discovered that, although the ectoplasm was normally invisible, it showed on the film as a whitish substance.

Sir William Barrett, a major founder of the SPR, was also in Ireland and visited the circle to investigate, accompanied by a very sceptical friend.

They were both impressed and delighted to find that sufficient light was available to observe what was taking place. They found themselves unable to catch floating objects. They also found it quite impossible to press down the levitated table. Barrett actually sat on the table which was not thereby pushed down to the floor; instead it tipped him off. Barrett read a paper at an SPR meeting later and Crawford wrote a number of books about it all.

The astonishing thing about all this was the attitude of the leading members of the SPR to it. Mrs Sidgwick dismissed the phenomena with contempt and suggested that the medium's leg was more likely to be the cantilever than ectoplasm. It really is quite astonishing that such a highly intelligent lady could make such a silly remark! Another distinguished member of the Council, Sir Oliver Lodge, congratulated Crawford warmly and put questions which Crawford answered. It is sad for the SPR that, because of the attitude of certain senior SPR members, Crawford published his material in the journal *Light* instead of in the *SPR Proceedings*.

Taken at the turn of the century, this is a rare photograph of ectoplasm — a tenuous greyish substance which can assume a human shape.

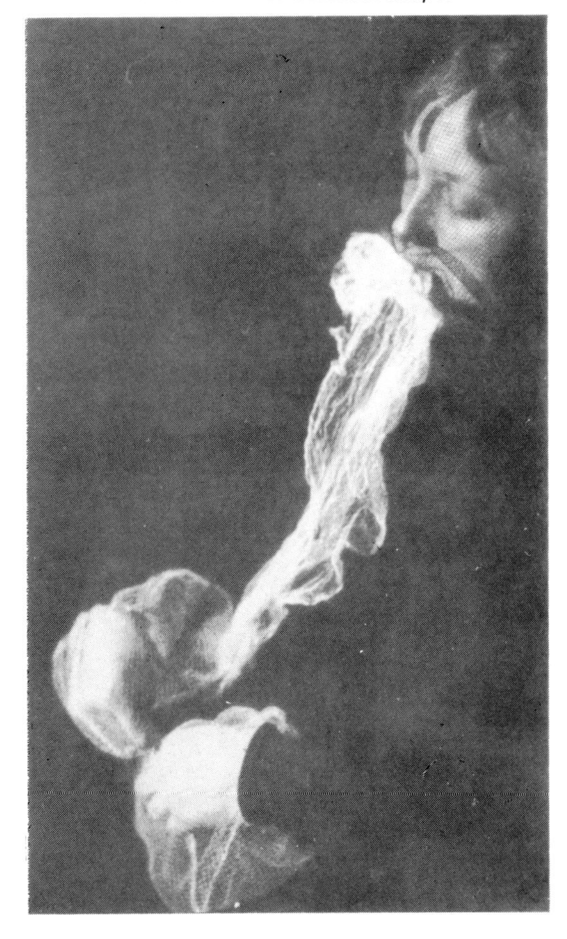

PERSONAL EXPERIENCES

It is difficult to see how the evidence for any phenomena could be better illustrated than in the cases I have so very briefly described. (Fuller and excellent descriptions are to be found in the books of Brian Inglis.) However, because of the materialistic climate prevailing at the time, few people accepted them. What are we to think today when, for some reason, the physical phenomena of mediumship are to be found exceedingly rarely, if at all?

In order to understand these occurrences better, I have been to untold trouble to have unusual experiences. Some years ago I was able to attend regular Sunday evening séances in London and travelled from Rugby to do so. This involved a long walk home sometimes as late as 2 am but these séances were certainly worthwhile as you will see.

The weekly séances were held in the home of a Spiritualist and he kindly allowed me before every séance, to inspect his room thoroughly. It was obviously not fitted with improper devices to produce fake phenomena. The medium had a number of different controls for the various types of phenomena. One control would give the 'sermon' at the beginning, another control would produce apports, and yet a third would produce the materializations. The sermon is not worthy of further remark but the apports that followed certainly are. A number of small objects appeared in the room and there was no way that I could see of proving that they had not been brought in in the normal way. However, the following interesting event once occurred.

With the red light still on, the medium, apparently in deep trance, walked out from his corner and approached the host of the circle who was sitting next to me. (All the sitters were Spiritualists except for me.) The medium held out his two hands, side by side, and I could see to my astonishment a pink glow over the top of his hands. It was rather like the glow seen in a Crookes' tube when an electrical discharge is allowed to take place in a gas but, for all my electrical engineering background, I could see no possible cause for this particular glow. As I gazed at this pink glow at a distance of about twenty-five centimetres (ten inches), I could see something apparently taking place within it. A rather misty-looking shape gradually solidified and dropped on to the out-

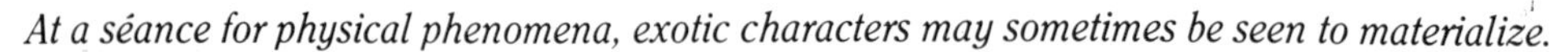

At a séance for physical phenomena, exotic characters may sometimes be seen to materialize.

stretched hands of the medium (which did not move from my sight during the whole episode). The object was a red rose which, the pink glow having disappeared, the medium picked up and handed to the host next to me. He thought that it was a rose presented to him by his wife from the next world. It is my belief that this occurrence, the whole taking place in a good light, was one of the most remarkable experiences I have ever had and, so far as I have ever been able to discover, appears to be unmatched by similar experiences of other investigators. Sadly I had no scientific witness to corroborate what happened, but my description is as accurate as I can give. (I have kept complete notes, extending over many years, of all the séances I have attended.)

Very often, at the beginning of these séances the medium would appear to be grasping large numbers of daffodils out of the air and scattering them around. I could not see where they were coming from and the whole procedure was a complete mystery to me until I had the experience described above. It is still a mystery but I can at least find it believable (and not as conjuring)!

After the session of apports the materialization control would then instruct the host to turn out the lights and cold breezes and materializations would commence. A long succession of different characters would apparently come out from the medium's chair, pick up the luminous plaques from the floor and, holding these next to their faces, then walk around the circle for inspection by us all. These characters ranged from Indians in turbans (one of them stated that he was the architect of the Taj Mahal) to Chinese, and Red Indians with babies and children! Occasionally we were allowed to touch the forms, which were cold and felt rather like solidified porridge (oatmeal). Sometimes it was possible to see the ectoplasm coming out of the medium's nose, mouth and ears and descending like a snake to the floor. I remember on one occasion a figure appeared in front of me who looked exactly like Sir Oliver Lodge, complete with big bald head and bushy beard. I hoped to have a discussion on physics or radio with him (during life he was a pioneer in radio) but sadly all he was able to produce was, 'God bless you, my son; keep on with the good work!' It was very obvious to me that whatever that form was, there was not much of Sir Oliver Lodge in it.

At the end of the séance — which often lasted three or more hours — the medium would be helped exhausted from the room. Again, what are we to make of all this?

First, there is no doubt in my mind that I observed that apport appear. However, I tried to persuade a very distinguished parapsychologist (who did not know me) to attend a séance and was completely unable to do so. It is only now, a great many years later, that I understand why. The reason is that parapsychologists are often regaled with 'facts' from people inexperienced in psychic phenomena which, although at first hearing are interesting, have a way of evaporating into normality on investigation. (I have often, in my lifetime, had that experience.) In addition, this particular parapsychologist was well-known to be an inhibitor of phenomenon (see page 136) and probably had few, if any, interesting paranormal experiences.

Let us have a few words about ectoplasm. It was the French scientist Richet who invented this word, describing it as 'a sort of liquid paste or jelly which emerges from the mouth or the breast ... which organises itself by degrees, acquiring the shape of a face or a limb'. Richet saw materializations in all stages from rudimentary shapes to complete forms and faces — some of them very imperfect and looking more like flat images than bodies. Occasionally the materialization was perfect. I have also seen ectoplasm behave in this sort of way. An important point I noticed, in regard to fraud, was that the ectoplasm I experienced so often had a very distinctive smell, rather like the atmosphere inside the carriages of a very crowded underground train on a wet hot day in the summer! If all ectoplasm is like this it would not be easy to imitate it!

An important feature of ectoplasm appears to be that it is 'idioplastic' (malleable by thought, the thought being the activities of George or a number of Georges in collaboration in a séance). I doubt very much whether characters of the next world have much to do with it — at least, only very rarely.

I noticed sometimes that the materializations, though, I feel, quite genuine, were apparently modelled on the face of the medium who was wandering around the circle. This is referred to by Spiritualists as matrix materialization and presumably occurs when the necessary resources for full-scale independent materialization are not present. I have no idea what these resources are and Spiritualists refer to them as the power.

The final point: when a materialization is taking place and ectoplasm has been exuded from the medium, who is normally in deep trance, if the lights are turned on or the materialized figure is violently interfered with in some way, then the ectoplasm is said to withdraw violently, still in its semi-tangible state, into the medium's body. This leaves bruising and the medium will be in a state of severe shock, verging on breakdown. Experienced sitters do not therefore touch the ectoplasm unless they are given permission by the control. The possibilities of fraudulence are thereby necessarily increased but that, unfortunately, is unavoidable.

A PRACTICAL APPROACH

The reader should remember that here in the West we have been conditioned by our culture and education to accept 'scientific materialism'. Just about all the paranormal phenomena with which the book deals are, according to this philosophy, impossible — therefore they cannot happen; therefore they do not happen. Our very language is based on the dualist hypothesis that we are the body and that all the objects of the universe are distributed around us in space. The body makes contact with these objects via the five senses and there is no other way of acquiring information. You will have gathered by now that I believe this philosophy to be fundamentally flawed. However, in order to break through this conditioning and approach the facts of the paranormal with a truly open and unbiased mind, it is necessary to have — at least I have found it so — first-hand experience. We find it quite impossible to fit into our concepts facts concerning the paranormal without personal experience of them.

What I usually suggest to people who are truly scientific and wish to take a scientific approach to these claims is to start off by having a few sittings with a really good medium or two. There are many Spiritualist organizations in London and the provinces, and in the United States, as well as many free-lance psychics, with whom appointments can be made at very reasonable fees. New sitters approaching the subject from scientific curiosity may not discover very good evidence for human survival of bodily death (mediums explain that they are in business to prove survival) but if the new sitters are lucky they will receive evidence of extra-sensory perception which will cause them to sit back in their chairs! It is suggested that several sittings should be arranged because sometimes there is no rapport with the psychic and absolutely nothing of value is available. (Good psychics in such cases normally return the fee to the sitter.) So give it a fair chance!

A new sitter, keeping an appointment with a psychic, should remember certain rules. It is usually a good idea to make the appointment over the telephone using a pseudonym, and to take a tape recorder (or pad and pencil) and record every word spoken by both of you. The reason for the latter is that closed-minded scientists, having had no experience, suggest that evidential information given to you by a medium is merely information which the medium has extracted from you earlier being fed back later. Of course this is far from true but new sitters should make sure that, while behaving in a friendly encouraging way they give absolutely no information away. The medium may ask them whether they understand certain information which he or she is passing as from a discarnate communicator; if the sitters do not understand, then they should say that they will think about it and perhaps it will fit into place later. Above all, do not return the medium's statements and requests as to whether you understand with blank negatives.

When you have obtained your notes (most sittings last about an hour and lead to perhaps between one hundred and fifty and two hundred separate statements made by the medium) I suggest you write each one down separately and divide them into columns labelled Correct, Incorrect and Not Understood. You will then be able to see what proportion of the statements made are correct and use your own judgement as to whether this shows evidence of something paranormal.

If you are really interested in making a scientific comparison why not persuade an acquaintance about whom you know very little to have a pseudo sitting, with you as the medium; make guesses as to

Medium / SPIRIT	Comments
Father HE LOOKS YOUNGER THAN ME!	His father was quite young when he died.
He mentions a photograph of his father with a moustache.	Correct
WE RESEMBLE EACH OTHER.	Correct
IT WAS SUCH A PLEASURE TO SEE HIM AGAIN. BUT HE DID NOT COME TO MEET ME.	
He loved music - good music.	Correct
There is music in the corner of the room, lovely music.	His tape recorder is in the corner of Diana's bedroom.
YOU WILL KEEP THE MUSIC GOING WON'T YOU BECAUSE THROUGH THE MUSIC I REACH YOU MORE EASILY.	
I KNOW YOU CAN THINK ABOUT ME NOW WITHOUT WONDERING WHY DID IT HAPPEN.	Correct
One of the girls had a birthday.	Correct
MAY I SAY HAPPY BIRTHDAY.	
Clothes in the wardrobe.	Correct
THERE IS A DINNER SUIT AND THREE EVENING SHIRTS. SEE IF SOMEONE CAN USE THEM. DON'T LEAVE THEM THERE.	Correct. We recently looked at the suit. Typical attitude.
SEE TO THE APPLE TREES. THEY WERE GOOD TREES.	Correct
WE HAVE GOT A LOVELY GARDEN.	Correct
Somebody had a walnut tree.	Not understood (In Mr Jones' garden. He had conversations with Mr Jones about this tree.)
Somebody learning to cook well, cordon bleu cookery.	His sister is so doing.
Betty.	Not understood
I DON'T WANT ANY PARSONS COMFORTING MY FAMILY	The parson came to see us shortly after he died.

possible communicators and suggest details of them and the life of the sitter. Allow the 'sitting' to occupy about the same time as the sitting you had with the medium, and ask your acquaintance to tell you how many of your statements were Correct, Incorrect, or Not Understood. This will give you some sort of assessment of how chance answers compare with the medium's score.

Of course there are better ways of doing experiments on ESP than this but it will provide you with a good first experience of the subject. You may be very surprised indeed at the result. But do remember that you are not conducting a test on survival but trying to discover whether it is possible for another human being to make statements about you and your life and about friends and relatives, alive or deceased, which they could not possibly normally know.

A final thought: If your score of correct statements compares very favourably with the psychic's perhaps it will be an indication that you have psychic capabilities. If this is so, in my experience, it is a little unlikely that you would not already suspect as much, from other indications.

To assess a sitting properly, it is vital to take clear notes of all that is said. This is one page of the notes taken during the sitting described on page 41.

5

PSYCHOKINESIS AND POLTERGEISTS

When I suggested — no doubt in company with others — that the BBC should bring Uri Geller to the UK from the United States, I hardly knew what I was starting! A great many remarkable and enlightening phenomena resulted. I remember so well that television *Dimbleby Talk-In* when Uri Geller demonstrated various phenomena which he claimed to be paranormal. The one which is perhaps best known and most strongly associated with him is 'metal bending' where he ostensibly gently rubs a normal spoon and it appears to become soft and plastic and bends under its own weight.

In fact, the incident that evening which astonished me most was the apparent convincing of a certain scientific academic of its paranormality — even though the conditions were far from what we usually refer to as 'test conditions'. There had been negligible control beforehand of the various spoons and other tableware which had been placed on the table for Geller's attention and the whole business was done in a television studio with no scientific control whatsoever.

Let me describe an experience of my own concerning Uri Geller. He was in the laboratory of Professor John Hasted at Birkbeck College (the University of London) with Professor Hasted (a professor of experimental physics), Arthur Koestler, Arthur C. Clarke and myself. Towards the beginning of the afternoon Arthur C. Clarke took from his key ring his Yale key and placed it on the back of the frame of the typewriter belonging to Professor Hasted's secretary. He placed his finger on the round end and the key was held flat to the metal surface. With his finger still on the key, Arthur C. Clarke then asked Geller to see whether he could do something with the key in that position, while Hasted, Koestler and I watched carefully. Geller gently stroked the key, moving his finger from Clarke's finger towards the end of the key; in about one minute the end of the key had curled up and this was made very obvious by Clarke's rocking it to and fro on the flat back of the typewriter with a finger at each end. There is no doubt in my mind whatsoever that this occurred: the key did not fall on the floor, Geller was not allowed to hold it behind him, he did not distract our attention; in fact the key did not go out of our sight in any way and four very experienced people watched it closely the whole time.

To those people for whom everything is black or white that episode would prove that Geller was genuine. However, Geller is an entertainer who receives large fees for his entertainments and, as

Children seem to be rather better at paranormal metal bending than do their more sceptical elders.

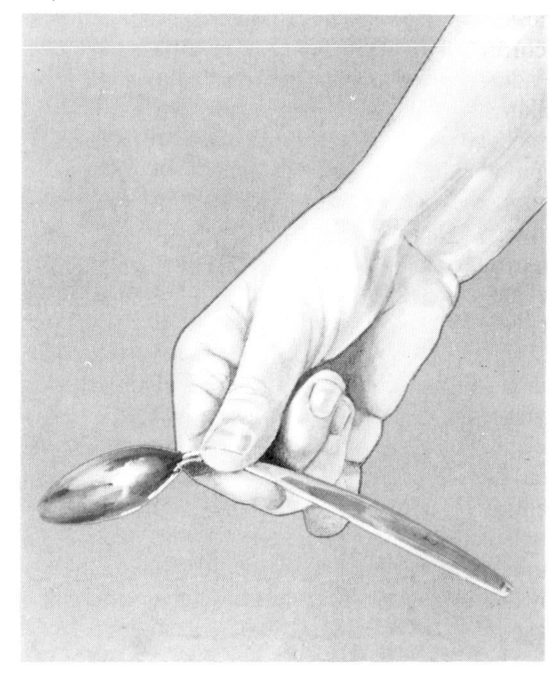

experienced parapsychologists well know, psychic functioning cannot be turned on like an electric light at will. So there is a very large temptation to help the entertainment out with some conjuring if the genuine article does not appear. I have heard that such 'normal conjuring' has been detected on occasion, but I have no firsthand experience of it in connection with Geller.

The most interesting part, to my mind, of Geller's activities was the large number of others, particularly children, who appeared able to carry out paranormal metal bending. Geller used to suggest over the radio and television that people should try to carry out an experiment with a spoon while he was doing it himself and this appears very often to have worked, the listeners and viewers finding to their astonishment that sometimes spoons did indeed bend paranormally. It is interesting that the same phenomenon appears to have occurred even though Geller's act was actually at that time being broadcast from a recording. His own consciousness did not appear to be necessary at the time. The perspicacious reader will perhaps realize that the beliefs of the benders involved were of great importance.

Arthur Koestler and I were receiving letters from a concerned mother in the north of England in whose home strange phenomena appeared to be occurring. These phenomena, based on Geller's broadcasts, were associated with two of her three children. She was worried about them and concerned for their health but particularly anxious that they should not fall into the hands of the media, who would exploit them. Arthur Koestler and I were convinced by this mother's honesty and undertook a journey further north to see the phenomena at first hand and, hopefully, to help if we could. Arthur, Mrs Koestler and I went to the home of the family first thing in the morning and on the way I bought a number of very heavy and substantial teaspoons from a shop. After meeting the family and talking to the children (a girl aged eleven and a boy aged thirteen) we did the first experiment. (I should perhaps mention to complete the picture, that there was a much younger sister who was not really involved.) I handed a teaspoon to the girl and another one to the boy. Arthur Koestler carefully watched the boy's teaspoon and I watched that of the girl. While we talked, with the father and mother of the children

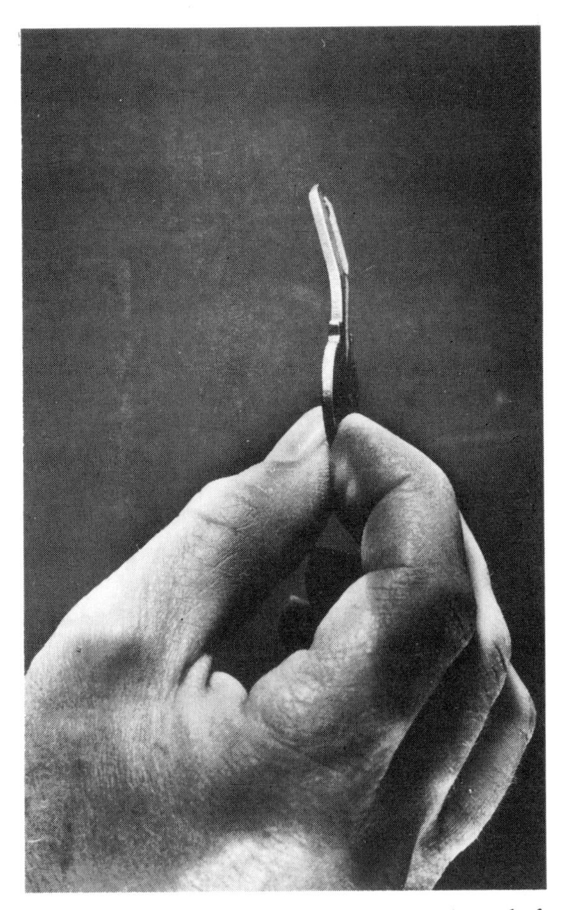

Experienced scientists witnessed the way the end of the key curled up when stroked.

looking on, the boy and girl each held their own teaspoon just below the bowl between the right-hand finger and thumb, the other three fingers being kept well out of the way. They gently rubbed the spoons between finger and thumb while we talked and watched. After about a minute, one of the two spoons appeared to become plastic and bent through about 120 degrees. There is not the slightest doubt about this: all occurred exactly as I have described and, as in the case of the experiment with Geller, the spoons did not fall on the floor, our attention was not distracted, and the spoons were in sight the whole time. And they were rigid again when I examined them closely. There was no way of telling that the bend had not been made 'normally' (by sheer force).

The newspapers were full of theories from so-called 'scientists' as to how such metal-bending phenomena were carried out. The theories ranged from the plain ridiculous to the downright absurd. I shall restrict myself to one particular fact. One of the broken spoons (broken by Geller) was sent in turn to two crystallography laboratories, one in Cambridge and one in London. One of those crystallography laboratories said that the fracture was very interesting and unusual and they would like to know how it had occurred. The other laboratory said that there was nothing notable about the fracture and wondered why we had sent it to them. That experiment did not tell us much about paranormal metal bending but it certainly told us a great deal about crystallographers!

Since those days Professor Hasted has been carrying out experiments on paranormal metal bending which is achieved at a distance. His specimens are held in mid-air, with attachments designed to register and make available for amplification the small movements which are evidently produced by a 'metal bender' sitting at a distance (this was non-contact metal bending). Hasted claims good results and has given many lectures and produced a book on the subject.

An experiment in psychokinesis

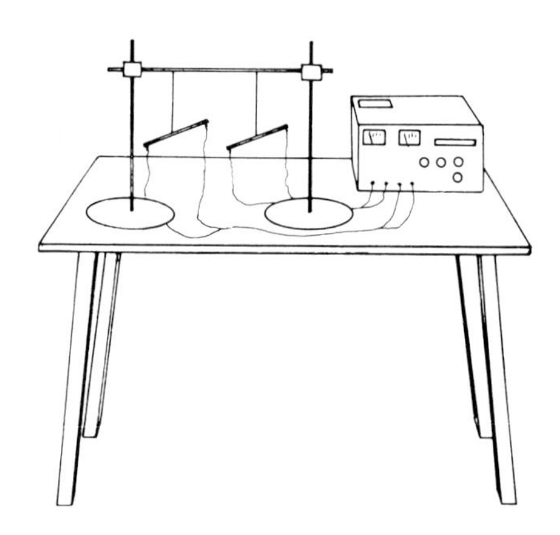

EARLIER WORK

Some of the earliest good quality scientific work on psychokinesis was carried out by Dr J.B. Rhine, working in the Psychology Department at Duke University, North Carolina. The term psychokinesis (PK) was actually introduced by J.B. Rhine to refer to the direct influence of mind on physical matter without the involvement of any known physical activity. His experiments involved random dice throwing from a machine, with the subject sitting beside the table and 'willing' the die to show a particular face when it came to rest. Rhine experimented for several years before he considered that his results were sufficiently definite to publish them. Some subjects could show very high odds against chance being the explanation of the results; in other words they were able to produce, with very much greater than chance frequency, the faces of the dice on which they were concentrating. There was the usual furore from scientists who, although they had no experience of the subject, felt qualified to pronounce that it was impossible. The Rhine results have since been repeated successfully in a fair number of other laboratories.

There have been many other claims of success in experiments involving different types of psychokinesis. The famous chemist and physicist, Crookes, long ago produced instrumental evidence that a certain medium could produce paranormal movements in a lever, its position being recorded on smoked paper. More recently, Kulagina, a Russian woman, has been observed producing movements of small objects like cigarettes on a table (but not in a laboratory).

I must mention one personal experience, again involving Uri Geller. Professor Hasted had arranged for an experiment involving a ship's compass and Geller was asked to attempt to move the needle at a distance. He was able successfully to do this and it rotated by what appeared to be some ten or fifteen degrees. The sceptical scientists will say that, even though the whole experiment was being recorded on videotape and Geller most certainly did not touch the compass, nonetheless he must have had a palmed magnet. In fact, that possibility was precluded by a *'Gaussmeter'* arranged near the compass which would have detected any change in the magnetic field produced by a nearby magnet. It indicated no such change.

However, many supposed cases of psychokinesis from a discarnate entity do prove to be nothing of the kind. For example, a lady telephoned the SPR office and informed us that each night, after she climbed into bed, her deceased husband switched off the bedside lamp for her. An investigator was permitted to sit in the bedroom and observed the lady turn on her bedside lamp and slip into bed. Sure enough after a few minutes the lamp went off. Sadly for the lady's comforting belief, careful examination of the base of the lamp indicated that the husband had connected in circuit a thermal switch which automatically turned off the lamp after a few minutes. The history of parapsychology is strewn with such misinterpreted normal phenomena! Misinterpretations and confusions abound, particularly in the minds of some of those who are not scientifically trained and also in the minds of a great many who are! *It is important to realize that expertise in other branches of science (or in magic) does not make an expert parapsychologist.*

Nina Kulagina, a Russian woman, is here shown moving a matchbox by psychokinesis. Such feats exhausted Kulagina and seemed to require great physical effort on her part. (Not all those skilled in psychokinesis react in this way.)

THE RANDOM-EVENT EXPERIMENTS OF DEAN ROBERT JAHN

Let us now turn to some very recent experiments that are of enormous importance with regard to psychokinesis. These have been carried out in the laboratory of Dean Robert Jahn of Princeton University (who was mentioned on page 30 in connection with 'distant viewing'). Jahn refers to the experiments as a study of low-level PK. He does not normally use psychics for his experiments but ordinary volunteers. Jahn's psychokinesis experiment uses a 'random-event generator'. It will not be necessary for us to understand the details but, in principle, random electrical noise (like that crackling sound that you sometimes hear on the radio when listening to a distant station) is used to produce a series of short electrical pulses. The pulses are then sampled — that is, the number of pulses occurring in a chosen short period of time is determined and this count is made for a chosen number of times each second. A certain number of pulses will tend to recur most frequently and quantities above and below this will occur less often, with very large and very small numbers of pulses hardly ever occurring at all. Anyone familiar with statistics will appreciate and expect that the distribution will be 'normal' or *'Gaussian'*, which is to be expected for such a random process and is well understood by any scientist. The reader not familiar with statistics need not be unduly concerned: all that matters is that the numbers of pulses are being counted. The pulses as they come out of the random-event generator equipment are automatically counted and the counts inserted immediately into a computer which makes all the calculations. The results of the counts are displayed on a visual display unit. They are also printed on a tape.

The experimental subject sits in front of the apparatus and attempts to make the pulses come either faster, or slower, or merely stay the same. The subjects use whatever mental strategy they like to achieve this. Some subjects like one set of ambient conditions which might be quiet music and dimmed lights; another subject might wish to go into a deep state of meditation. The conditions in the room and the mental activities of the subjects are chosen according to preference. Their success or failure, in speeding up or slowing down the pulses, is indicated on the screen. It is important that the computer causes the pulses to be produced and instructs the subject for each experiment whether to speed up the pulses, slow them down or do nothing. This is vital because if there is any 'drift' in the apparatus the rate of pulse production will be slowly changing anyway. By attempting alternately to speed up and then slow down the pulses, the risk of being misled by 'drift' is removed because any apparent success in one direction will be cancelled out by apparent failure in the other direction. It is found that many subjects (and, as already mentioned, Jahn does not use psychics) are able to achieve success in an experiment of this kind. Their degree of success is measured in terms of the odds against chance as accounting for it. Any speeding up or slowing down of the counts would occur by chance once every so often. If such an effect would occur naturally only in every few million experiments then a statistician would say that to succeed much more frequently becomes exceedingly significant: a high degree of success would indicate that the mental activity of the subject — the only important variable — had indeed altered the rate of production of the pulses.

The success of some subjects in altering the rate of production of the pulses is at a very low level; they are speeded up or slowed down by, approximately, only a half of one per cent. (The normal distribution is shifted to this extent.) However if this can be continued for a great number of experiments then the odds against chance accounting for it can be very high indeed. If the subject achieves nothing then normally the number of counts will wander around a 'base line'. Occasionally it will deviate from it by perhaps five per cent. However, if the subject is successful — and many have been — at causing the counts to speed up then they will deviate in one direction from the base line by very much more than this and this trend will be continued the longer the experiment continues. Jahn has achieved odds of many millions to one against chance as accounting for this. This is very impressive evidence to support a claim about the existence of psychokinesis.

Jahn has also noticed a most interesting and intriguing phenomenon. When some subjects try to speed up the pulses they in fact slow down and, conversely, when they try to slow them down the pulses speed up. Perhaps their George has some sort of perverted desire to obscure the fact that psychokinesis is possible and deliberately overdoes it in the other direction — proving just as definitely as before that it occurs! Some people, it is found, are able to change the speed at which the pulses are produced in only one direction (still with enormous odds against chance as accounting for it).

This experiment is continuously and automatically checked as it proceeds and if anything at all goes wrong with the counting or in any other respect, it is immediately detected and the experiment disregarded.

Certainly, different subjects appear to be able to do different things, some being able to achieve a cumulative deviation from chance in one direction but not the other, others perversely achieving the opposite direction from that aimed at. Jahn refers to these characteristics as the 'signatures' of the subjects concerned.

Jahn stated, several years ago, that when all his formal data were put together, the cumulative deviation gave odds of one thousand million (or more) to one against chance as accounting for the effect.

Several questions naturally would arise in the mind of a good scientist when hearing of those results. The first question would be, 'Is the effect specific to that particular microelectronic noise source?' Jahn changed the noise source and it made no difference. He even built a complete second model of the equipment but again this made no difference to the results.

The next question to arise would be, 'Is the effect observable macroscopically as well as microscopically?' (in the large as well as in the exceedingly minute). No one has ever seen an electron and it is this which is at the root of the electrical noise source. Jahn therefore built an entirely different kind of equipment demonstrating a random process; he called this a 'random mechanical cascade'. It was a vertical panel some three metres (ten feet) tall to which were mounted over three hundred horizontal pegs. Nine thousand lightweight spheres, each a little less than two centimetres ($\frac{3}{4}$ inch) in diameter, were allowed to descend from a hopper at the centre of the top of the equipment and trickle down through the pegs – rather like a pinball machine. They would bounce on the various pegs, strike against each other and eventually all would reach the bottom of the board. At the bottom he arranged nineteen collecting bins, tall thin containers, to collect the spheres when they reached that level. As one would expect, more spheres enter the centre bin (the one immediately under the opening at the top of the board) than enter the bins on each side. Not many balls reach the extreme left-hand or right-hand bin and the 'distribution' is the same sort of curve as was described in connection with the electrical noise experiment, what a statistician would call normal or Gaussian. (In my first university we used equipment rather like this to demonstrate what a random process is like to first-year electrical engineering students.)

Jahn's 'cascade' incorporates a light-beam detector which counts the spheres as they enter each bin, and inserts the number into the computer. The operator sits some three metres (ten feet) away and attempts, either by choice or on instruction, to

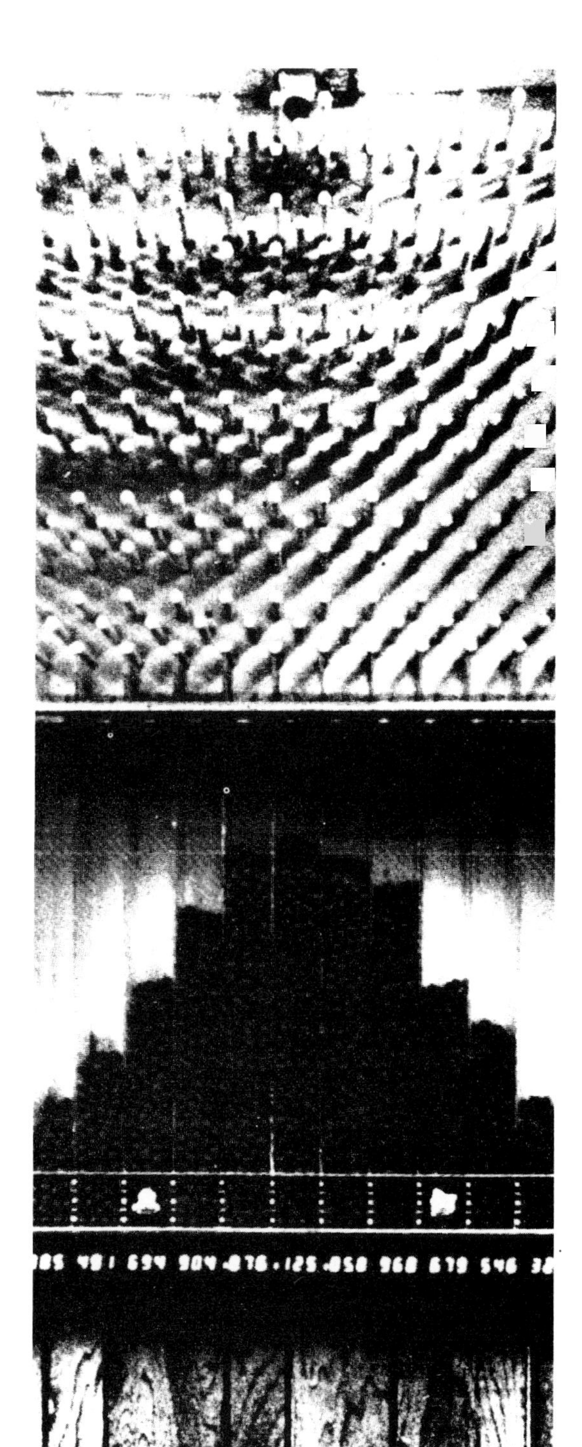

Jahn's random mechanical cascade is an interesting piece of equipment for examining P. K.

cause the spheres to move either to the left or to the right as they descend; in other words, he or she tries to distort the distribution in that direction (just as in the electrical random-event generator experiment). All the data are acquired in sets of three runs in which attempts are made to cause the balls to move first to the left, and then to the right and then in no particular direction. This variation is again necessary in order to counteract the effects of any 'normal' external influences (such as the board's not being precisely vertical or perhaps draughts entering under the screen and thus causing balls to move in one direction or the other).

In this 'random mechanical cascade' experiment the accumulated deviation of the mean bin number is calculated for each side separately, the base line being the normal distribution when no mental effort is being exerted.

The results of this experiment are remarkably similar to those of the random-event generator experiment. Efforts to distort the mean bin number to left or right again often progress in the intended direction — in a way very similar to that of the results of the other experiments and, most remarkably, the 'signature' of a given operator is found to be the same. What this means is that operators who found they were able to shift the distribution to the right, that is, they were able to speed up the electrical pulse counts in the first experiment, were able to shift the mean number of balls to the right in the second experiment. Those who could not slow the counts down, that is, to shift the distribution to the left in the other experiment, also found they could not cause the average number of spheres to move to the left.

An interesting fact was observed during one experiment when a subject was having success in causing the balls to move to the right (the scores being indicated on the visual display as before). It became apparent that the results went back to chance again (returned to the base line) when the operator elected to try a series without the LED displays of the bin populations being illuminated. He clearly needed this feedback of information in order to succeed.

In an experiment of this kind it takes several minutes for all the spheres to fall to the bottom of the equipment so it is clearly not possible to build up such enormous odds against chance as it is in the random-event generator experiment; an equivalent large number of experiments simply cannot be completed in the time available. However, Jahn stated that if the entire data base from eight operators who had completed at least one series of experiments was compounded into one record then the 'statistical aberration was significant to the order of 10 to the power minus 4'. In other words, there would be one chance in ten thousand that the result had been achieved by chance. Put another way, if the whole series of experiments had been repeated ten thousand times then that result might have been achieved only once. This is again highly significant to a statistician.

How can these data be interpreted meaningfully? The first fact of importance is that the psychokinetic effect is of a particular kind depending on the subject: it is 'operator specific'. It is also often 'condition specific': it depends on the ambience amongst other things. It is not however nearly so 'device specific': that is, it does not depend on whether the random process being affected by a mental agency only is produced by electrical noise or by falling spheres.

It looks as though one should not be tempted to represent these facts by a mental model of the human mind 'pushing objects' in the world out there (which is implied by the very word psychokinesis). Something very much more fundamental is clearly involved and Jahn suggests that this is 'the degree of information implicit in the distributions'. Jahn suggests that a more fundamental model is needed than is provided by any one of the current possible scientific models. Under the auspices of his present research programme, efforts are being made to model the facts using electromagnetic models, thermodynamic and mechanical models, statistical mechanical models, hyperspace models, quantum mechanical models and others.

We have to consider what we mean by 'reality', 'consciousness' and 'environment'. Jahn suggests (and many others have, of course, done the same) that reality is constituted only when a consciousness interacts with an environment and neither can be legitimately described in isolation. The physicists working in quantum mechanics use just that sort of language. Reality is an 'interface between two domains' and Jahn suggests that the currency of that interface is information, and that information can flow in either direction. He suggests that the existence of psychokinesis implies that the consciousness is inserting information into these random physical systems. By the use of the mind the subject is altering the 'description' of what is occurring. We are used to the idea of the senses informing the mind of the description of what is 'out there'; we are not yet used to the idea that the mind itself can alter that description.

There are certain most important deductions which it is reasonable to make from these magnificent experiments. Let us consider just one.

More and more important processes are being controlled by microprocessors these days; large computational systems are in control of many aspects of our lives — such as factory processes, aircraft and space probes. These microprocessors are arranging switching at less and less energy as they become more evolved. The amount of energy is perhaps approaching that of the order of the electrical noise in the random-event generator experiment. Imagine what might happen to these controlled systems if someone who could achieve exceptionally good scores with the random-event generator attempted to disrupt the process or equipment so controlled. And they might do it unconsciously! It is a rather frightening thought! One hopes that computer manufacturers will make themselves aware of these facts and support scientific research in psychokinesis. I must confess that at the moment I do not see how the possibilities can be completely avoided. But who knows!

POLTERGEIST PHENOMENA

The word poltergeist, as most people know, means a noisy boisterous ghost. Objects are reported unaccountably to fly about, especially stones, and sometimes they are said to follow curving paths around corners. Poltergeist phenomena have been reported as the causes of various items of damage, spontaneous fires starting, water unaccountably appearing, mocking laughter, groans and screams, and people being touched or smeared with unpleasant substances but usually not otherwise hurt. Because poltergeist phenomena seem to be rather like psychokinesis in some respects, they are usually referred to in the technical literature as recurrent spontaneous psychokinesis, abbreviated to RSPK.

A very well-known and fairly recent case occurred in a lawyer's office in Rosenheim, West Germany, in the late 1960s. This case involved phenomena which were exceedingly well evidenced. Pictures swung round and lamps swayed violently; all this was filmed. In addition, electric lamps exploded and fuses unaccountably blew. Movements of objects and furniture were recorded and there were many calls by telephone to the speaking clock — in fact many more calls than it would have been possible to dial in the time. The electric-power company installed a recording voltmeter to measure the variations in voltage in an attempt to account for the lamps and fuses failing and they found a number of large variations in voltage at irregular intervals. It was noticed that there were loud bangs at about the same time as the voltage deflections but not necessarily one bang for each deflection.

The Rosenheim case was resolved when it was discovered that all the phenomena occurred when a certain nineteen-year-old secretary came into the office. When she went away the phenomena ceased. She is now happily married with children and, so far as I know, has experienced no further episodes of this kind.

It is clear that poltergeist phenomena are usually linked to one particular person. This is in contradistinction to haunting-type apparitions — which appear to be linked with one particular location. Usually, but not always, the particular person (called the focus) is an adolescent child and generally he or she is undergoing some sort of stress. A fair proportion of such subjects are found to have the symptoms of epilepsy or physiological and physical conditions which may lead to epilepsy.

Poltergeists usually appear to be very shy of investigators! Often the phenomena cease as the investigators come through the front door and start again when they leave. The uninformed would therefore assume that they are being faked by the family involved but this is by no means a reasonable assumption. Sometimes children in poltergeist cases have been caught deliberately helping the phenomena, perhaps by throwing something when they thought the investigator was not looking. Probably such children, who are often under stress, like being a centre of attention and do not wish the phenomena to cease. That there is evidence of cheating does not mean we can easily dispose of all poltergeist phenomena. The evidence that they do in fact occur is far too good.

It is not often possible to observe this but it has been noticed in a few cases that the phenomena appear to be stronger near the focus and weaker further away. Probably more than half the foci seem to be adolescent children but this by no means applies to all the cases. Also, a state of psychological stress is often discovered but, again, there appear to be cases where this is not evident. Often it is as though the poltergeist phenomena are in some way using the pent-up mental energy of the stress by converting it into a physical form of energy.

The well-known English psychic Matthew Manning was as a schoolboy surrounded by poltergeist phenomena at home and at his boarding school. These did not decrease in intensity until he found other ways of using his psychic energies. In this case Matthew Manning appeared to be psychic but, so far as I know, no more stressed than most schoolboys of his age.

I have not as yet been fortunate enough to be present when poltergeist phenomena have actually occurred. However, I have certainly observed the paranormal movements of objects in séances. I was involved in one ostensible case of poltergeist phenomena when water was unaccountably appearing on the beds in a cottage and objects were described as flying about. I was able to inspect the damage done but did not witness its actual occurrence. The witnesses were fairly numerous but there were no first-hand witnesses having an adequate knowledge of science for critical assessment. The phenomena here were complicated by unusual electrical conditions which might well have provided a normal explanation for some of the phenomena — such as the inadequate earthing.

A well-known British researcher, G.W. Lambert, put forward a theory to account for some poltergeist phenomena: his underground water theory. In a number of cases Lambert found evidence that poltergeist phenomena appeared to occur in houses beneath which underground water was flowing and were more frequent immediately after heavy rain-showers. He plotted the path of the River Fleet (near Fleet Street in London) under the famous Cock Lane where there were claimed poltergeist phenomena. He identified sites with the path of the underground stream. The gurgling of water in a culvert could well be mistaken for mocking laughter. No doubt water pouring through an underground pipe beneath a house would cause a certain amount of vibration but it does not appear that it would be sufficient to cause crockery to fall off shelves, let alone cause it to fly through the air. I do not believe that many parapsychologists would consider that Lambert's theory explained many cases.

Probably the solutions to these phenomena will one day be found. Some ideas about reality and the flow of information were briefly mentioned after the discussion of Jahn's psychokinesis experiments and they will be more fully discussed on pages 150. Poltergeist phenomena appear to be a way in which mental stress is resolved. This seems to point in the direction of the mind/universe interface. When this is understood perhaps the reasons for a great many of these strange phenomena will then become much clearer.

Over the years, a number of cases have been appearing of subjects who claim that their electrical equipment burns out considerably more often than is the case with all their friends. Also, a correlation appears to be taking shape between such people and allergies. It is too early yet to comment or draw conclusions but it is mentioned so that anyone who knows of any such cases can appreciate that it may be part of a pattern. The author would be grateful to be informed.

SEEQ: SEEK THE SEQUENCE

SEEQ is a psi-game in which the object is to match sequences of ten digits. In response to your success in doing this, a spot moves to and fro leaving a trace down the screen. In the abscence of psi, the trace will tend to oscillate about the centre line. The aim is to induce a right-hand bias into the trace, so that the spot eventually breaks through a 'wall' into a coloured zone. Auditory feedback is provided in this zone — the idea is to keep the tone sounding. The 'wall' and right-hand edge of the screen correspond to chance probabilities of 1 in 20, and 1 in 1000, respectively.

The game may be played in either an ESP mode or a PK mode. In the ESP mode, the computer generates a hidden (random) sequence of ten digits (1 to 5) which you then attempt to match. In the PK mode, the computer displays instead a pre-programmed pattern of ten digits, and you then attempt to influence the computer to generate a matching sequence.

Each game consists of from 10 to 100 trials, with optional trial-by-trial feedback as to which digits have scored hits. At the end of a game, the computer displays your scores and cumulative scores, plus various statistical information (including psi-coefficients, critical ratios and odds against chance). A number of players may compete for the highest score. At the end of a session — consisting of up to 25 games — an overall trace and results are displayed. The SPR will be interested to hear from you in the event of your achieving highly significant results!

A unique feature of the game is that it has an underlying *real-life psi-quest:* the solution of a mystery which has tantalized psi researchers for many years. In the 1940s Dr S.G. Soal carried out a famous series of telepathy experiments with outstandingly successful results. However, anomalies have been found in his target lists, suggesting that he almost certainly manipulated the data to produce spurious hits. Soal claimed that he derived his target lists, consisting of digits 1 to 5, from logarithm tables. If it were possible to trace the source of the target lists it would become clear to what extent Soal manipulated the data; it might even prove possible to clear his name. Unfortunately, all attempts at identifying any target sequences have failed. Nevertheless, with the help of the readers of this book, and the thousands of searches in the logarithm tables that the playing of SEEQ will make possible, the mystery could yet be solved. Will you be the one to make the breakthrough and find the missing key to the logarithm tables? A prize awaits you if you do, and your name may go down in the annals of psi research!

The program is written in BBC Basic. Cassette tapes, suitable for running on the ELECTRON and BBC microcomputers, are available at £3.50 each (including postage) from the address below. Alternatively, a copy of the listing, plus notes on modification to SPECTRUM Basic, may be supplied at a cost of £1 to cover photocopying and postage.

Betty Markwick
5 Thorncroft
Hornchurch
Essex RM11 1EU

6

OUT-OF-THE-BODY EXPERIENCES

TYPICAL EXPERIENCES

About one person in ten has one experience or more of perceiving an apparition during their lives. Much the same proportion of people have an experience of apparently leaving their body and many of them have the experience of actually 'seeing' it as from another point in space. If this occurs to those who have never heard of an out-of-the-body experience it is exceedingly alarming and sometimes leads them to think they have died. It is also quite alarming when it happens to those who have read about the experience; there is always the fear that they will not be able to 'get back'. Happily most of these fears appear to be groundless. I have never yet heard of anyone who came to much harm as a result of such an experience. (It may occur to the perceptive mind that if someone had the experience and was unable to return to the body then he or she would presumably be found later dead from heart failure or something of the kind. My feeling — and clearly I cannot prove it — is that there are probably no such cases.)

So what exactly is it like to have an out-of-the-body experience?

There are many ways of 'leaving the body'. (I have used quotation marks because, even though the experience is exactly as though one moves away from the body into another part of physical space, this is by no means a useful way of looking at it.) However, it is simpler for the moment to use that term and we shall explore the difficulties a little later, on page 75.

For some people the experience occurs spontaneously. They may or may not be lying down when they have a strange feeling, hear a sound like rushing wind or the sea (with or without a loud humming noise) and then they seem to float out of alignment with the body so that they have the experience of observing the room from a different position in space and not through the physical eyes. Other people are perhaps sitting quietly in a chair reading and they discover themselves in another part of the room observing their other 'physical' selves still reading. Yet others find that during a surgical operation with total anaesthesia, they are apparently floating in the air in the corner of the operating theatre looking down on their body and watching the team of surgeons and nurses who are performing the operation. Sometimes they overhear what is said and are able to report it accurately afterwards — to the astonishment of the members of the operating team. Now and then, during such an out-of-the-body experience, those taking part find they have a wide-ranging view of other parts of the building and the countryside around; occasionally they may appear to receive the thoughts of others.

During an OOBE, some people feel that they float out of the body horizontally through the head while others slip through the feet; sometimes they 'roll out' sideways and yet others float upwards horizontally. Sometimes they are cataleptic (unable to move a muscle immediately before the experience); sometimes they are not. Some see a sort of long greyish flexible pipe joining their two bodies; again others do not. Some see this so-called 'cord' joining the front of the head of the physical body to the back of the head of the 'other' body. Some see it joining the foreheads of both bodies. Yet others do not see anything like that at all. The Bible speaks of death as the 'golden bowl' being broken and the 'silver cord' being loosed. [Ecclesiastes 12, v.6.]

It is sometimes found that the centre of consciousness appears to go with the second body.

Sometimes the consciousness is dual — that is, the experiencer is partly conscious in the other body and partly still conscious in the physical body. Cases have been known where the two bodies have 'talked' to each other. Occasionally the experiencer in the 'other' body finds that he or she is partially or wholly 'somebody else'.

A certain proportion of hypnosis subjects can have an out-of-the-body experience as a result of suggestion under hypnosis. Others can not; suggestions that they will move out of location with their physical body have no effect. Why this should be so I have no idea.

There have been cases where only part of the other body is 'exteriorized': perhaps the top half sits up while the whole physical body remains lying down, the bottom halves of the two bodies remaining in alignment. There have been cases where the hand of the other body has been used to feel down through the carpet and floor boards and, as it were, explore what is in the space between the floor and the ceiling of the room below.

The experience of observing the room from an out-of-the-body position is by no means a simple one. Sometimes the room has modifications which may be symbolic. For example, there could be bars on the windows which prevent egress. There may be small but significant changes in the surroundings. I

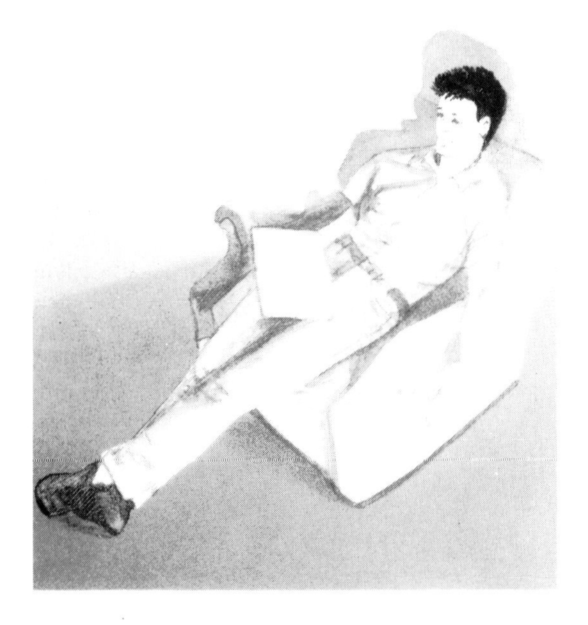

remember one fellow SPR member telling me that he had an out-of-the-body experience every time he fell asleep on his back. During one of these experiences he observed his laundry bag beside the dressing table, waiting to be dispatched. After returning to his body he noticed that it was not in fact there, having been collected earlier in the day. Sometimes everything is lit by a strange glow and often the room is apparently observed quite clearly although it is the middle of the night and dark. It is obvious that what is being observed is by no means the ordinary physical room but a sort of dramatized reconstruction of a memory of the room. This again is perhaps an oversimplification.

Occasionally people who are having this experience may try to move a small object, perhaps attempting to switch on a light. Usually they find that their fingers go right through the object. Sometimes, however, a small object is said to have been moved. Occasionally an experiencer has been seen by someone else. The out-of-the-body experience appears to be different from 'telepathic projection'. Remember the 'apparition' of Mr Beard in the middle of his experiment when he attempted to 'project' himself to the room of the Misses Verity in Kew

Some people feel that they float out of the body through the head. This other body may be nude but is not necessarily so.

(page 18). In this case Mr Beard was not conscious of moving through space or having an out-of-the-body experience at all. Although there may be similarities between the two experiences they are clearly not identical. However, the differences may be less than they superficially appear.

Let us mention one or two further points. Sometimes the other body is found to be dressed in night attire, just as the physical body was when it was 'left'. Sometimes it is found to be nude. Occasionally the other body is dressed in conventional indoor or outdoor clothes or even flowing Grecian-style robes. Perceptive percipients have discovered that they can change their clothing (and body!) at will by the conscious use of imagination.

Occasionally the surroundings of the physical world are found to alter dramatically in an unaccountable way and the experiencers find themselves in very beautiful countryside, with or without other people present.

There have been cases where evidence has been obtained from another place, checked later and found to be accurate. Sometimes such experiences are found to be fantasy.

THE NEW YORK EXPERIMENT OF EILEEN GARRETT

The famous psychic Mrs Eileen Garrett once took part in an experiment in which she had an out-of-the-body experience while surrounded by experimenters in New York City. She projected to a place in Newfoundland where another experimenter had arranged to collaborate with her. Her dual consciousness enabled her to describe the experiences of the projected body through the lips of the physical body on the couch in New York City. She referred to the tang of the nearby sea, and to the flowers near the path. She describes going into the house and seeing the other experimenter descending the stairs, his head bandaged. She reported all this back to the New York group. The experimenter in Canada appeared to be aware of her and went

Eileen followed the experimenter into his library

into his library, she following. He took a book from a shelf, opened a page at random and read some sentences. Eileen Garrett says that she could both see the book with the sentences he was reading and also pick up from him telepathically his thoughts as he read them. The experimenters in New York wrote all this down and posted it to the experimenter in Canada. At the same time the experimenter in Canada wrote down what he had done and posted the details to the experimenters in New York. Eileen Garrett says in her autobiography *My Life as a Search for the Meaning of Mediumship* that the two accounts tallied in every way.

Someone naïve to this subject would think that the explanation of this experiment was perfectly clear. Mrs Garrett had travelled to Canada in a subtle body, had observed all the facts which she had transmitted and had then returned to reintegrate with her normal physical body. However, this may not be the best way to interpret the experience. An alternative possibility is that Mrs Garrett's George had picked up by telepathy from the experimenter in Canada all the facts which she had transmitted and George had dramatized them to produce the apparently objective experiences she had. Mrs Garrett was a distinguished psychic and her George (see page 12) would presumably have been good at that sort of thing.

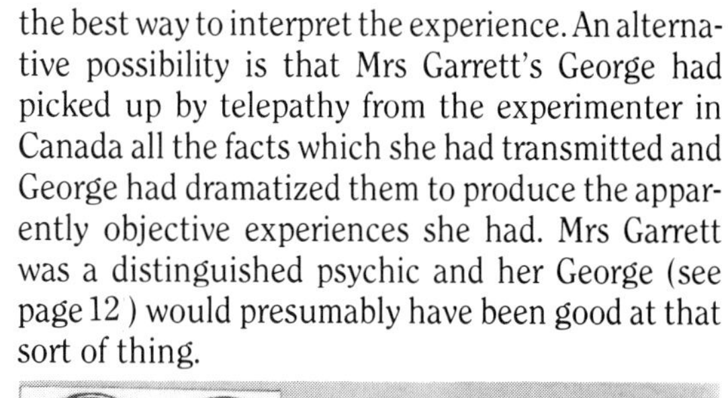

Eileen was able to describe her out-of-the-body experiences as they occurred.

OUT-OF-THE-BODY CLAIRVOYANCE

So is it possible in any way to determine facts about the physical world without involving one's own or anyone else's five senses? It is indeed and that amounts to a test of pure clairvoyance, such a test having been discussed earlier in chapter 3. For the purpose of making a test on whether information about the physical world could be acquired during an out-of-the-body experience I devised and had built the equipment described on page 28 . A colleague and I had earlier prepared a list of volunteer subjects for experiments using hypnosis. We had hypnotized all of them successively and carried out tests to find out whether they would have an out-of-the-body experience when it was suggested to them under hypnosis. Those who had an apparently satisfactory out-of-the-body experience we put on our list.

One of these subjects agreed to come and take part in the experiments. I carried out the necessary procedure — I hypnotized her and suggested that she would have the experience of moving away from her body but that she would still be able to talk to me through her lips. She had such an experience. I then pressed the button on the front of the box which produced a three-digit random number at the back. I asked her to look at the back from her other body and tell me through her normal lips what the number was. She gave me a three-digit random number. As this was a first and pilot experiment I looked at the numbers at the back of the box and discovered to my delight that the number she had stated was there displayed. So I pressed the button again and produced a second number which she gave me. Again there was an agreement. With mounting delight I then said, 'Now we will do a proper experiment and I shall not look at the numbers at the back', so I pressed the button at the front and asked her for the three numbers displayed at the back. She gave me numbers with some difficulty. (As I did not look — that would have vitiated the experiment in pure clairvoyance — I do not know whether they were correct or not.) I set them up at the front and pressed the button again for a second number. She then said, 'The numbers are awfully small: I am having great difficulty in reading them.' We had to discontinue the experiment and I asked her to go away and practise reading small numbers of the size displayed on the box. I first gave her instructions as to how she could use autohypnosis to produce her own out-of-the-body experiences and she was able to do this satisfactorily. She went away, as she explained, to ask her fiancé to produce small numbers and set them up in another room when she would attempt to see them through her out-of-the-body experience. Perhaps you can guess what happened! She diverted her energies into getting married, moved to another part of the country and never came back. A later experiment with another subject was not nearly so successful and I eventually abandoned the experiments.

Some months later a well-known psychic from the United States visited me. During conversation I told him about my random-number box and he immediately expressed a wish to try a run of tests. The results of this experiment in pure clairvoyance are described in chapter 3. You will perhaps remember that the very high score of eight in a run of twenty was due to bad contacts on a microchip and I very tentatively suggested that perhaps the psychic's George found it easier to spoil the electrical contacts than to acquire knowledge of the numbers by pure clairvoyance.

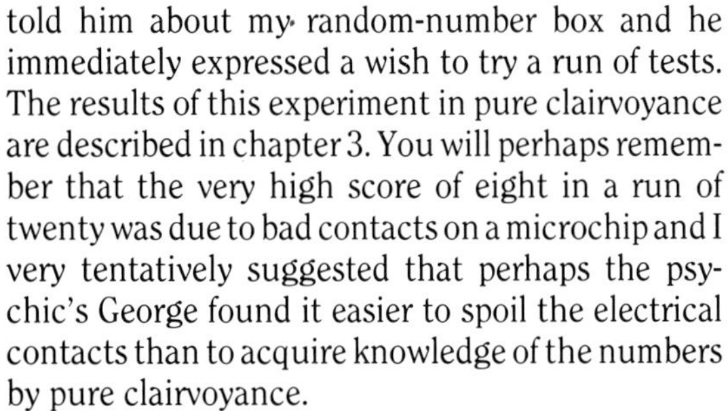

There have been a number of ingenious experiments made to determine whether it is possible to obtain knowledge of the physical world without anyone's five senses coming into the actual experiment. It will be appreciated that George is well able to create an appearance of the normal physical world from information he may have obtained telepathically. In the experiment of Dr Karlis Osis (of the American Society for Psychical Research) which I will now describe, a test is made to see if it is feasible for the consciousness of a percipient to have moved through physical space and in some way be receiving light rays from the surroundings (even though the physical eyes are absent).

The subject had an OOBE, during which she tried to read the numbers.

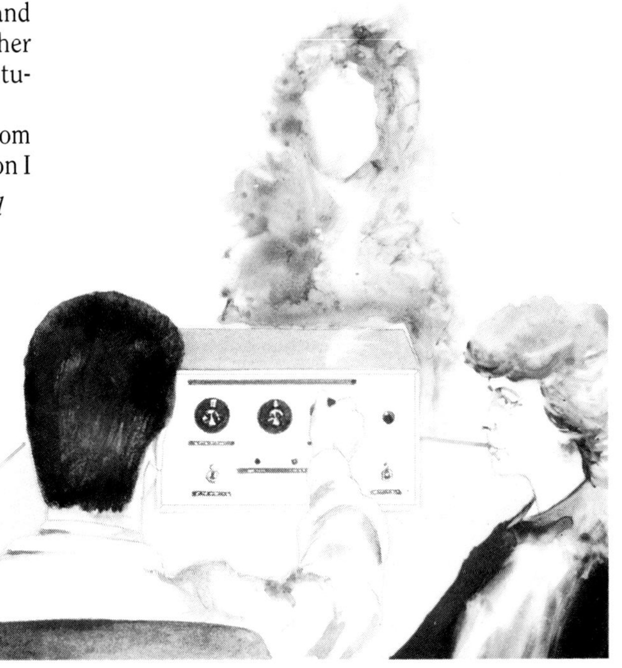

The experiment involved, in its simplest form, having a letter 'p' within a closed box. This letter was reflected in such a way that, if looked at through a window into the box, it would appear as the letter 'b'. The idea was that if the subject of the experiment saw the letter by ESP then he would see the 'p' whereas if he were in some sense looking through the peep-hole then he would see the letter 'b'. This idea was further developed to display different pictures in several colours and in four quadrants. The picture seen through the window of the box would be the sum total of black and white outlines, with colour added by a colour wheel and involving several mirrors. By ESP one could not, as it were, see a complete picture. This could be observed only by seeing the light coming out of the window. An American psychic (Dr Alex Tanous) doing this experiment lay down in a sound-proof room and was asked to have an out-of-the-body experience, go to the laboratory and look through the window, returning to describe what he had observed. The psychic actually had an out-of-the-body experience in which he did not seem to have a body but was rather a point of consciousness. It took him a little time to succeed but he did eventually have some success. Feedback was provided for him; that is he was told whether he was right or not. He did appear to be learning how to do it and the results, which involved seeing the complete picture, appeared to be approaching significance.

This experiment was analysed in rather complicated ways and produced no clear-cut or final answers. In any case this may be too much to expect as we do not know the limits of extrasensory perception and it might be perfectly possible in these circumstances to determine what the entire synthesized picture would be like by ESP. Certainly, as mentioned earlier, the results of psychic functioning appear to be 'goal directed' — understanding the details of equipment does not appear to be necessary. This is a very difficult and bewildering subject.

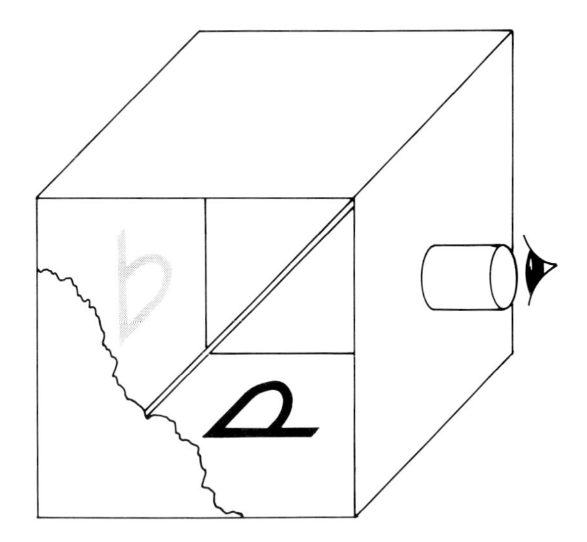

Many different experiments have been done in laboratories, using the out-of-the-body experience as a basis. In one, the experiencer, Dr Keith Harary, was asked to try to influence the movements of his cat towards a chosen sector of a particular marked circle. He appeared to have success with one cat but not with another; that is, one cat seemed to be in some way perceiving him and was influenced by him. However, one could not truly claim that the result was very significant! In another experiment Dr Charles Tart, an American parapsychologist, asked a young lady who had frequent out-of-body experiences to read a five-digit random number placed on a high shelf. She did this successfully.

I can hardly complete this very brief description of a few aspects of the out-of-the-body experience without describing my own.

THE AUTHOR'S EXPERIENCE

Some years ago I considered that if I could have an experience of being apparently separate from my physical body then I would have evidence which might help me to decide whether human beings survived bodily death. I proceeded in the usual scientific way. I read all the relevant books which had been written on the subject and then attempted my own experiments. One book, *The Projection of the Astral Body* by Muldoon and Carrington, was particularly helpful in that Muldoon described various methods he had found to be successful in giving him an out-of-body experience. (In fact, in those days the out-of-the-body experience was called astral projection, the assumption being that we had an astral body normally interpenetrating the physical, which we could move out of alignment and use to travel around physical space.) So I spent one hour in bed each night before I went to sleep practising the various exercises of imagination and 'willing' the movement of the astral body described in the book. They involved such things as going to bed thirsty and imagining oneself going to the kitchen for a drink of water. Another method involved imagining oneself looking at the ceiling while lying on one's back and then letting the vision move across the ceiling, down the wall, across the floor under the bed, up the other wall and back to the ceiling — without actually moving the physical body at all. There were various other schemes described and I spent one hour a night trying each of these in turn.

One month after commencing these exercises my interesting experiences began. Without any break in consciousness and with the light in the bedroom still on but my eyes closed, I found myself in a cataleptic state — that is, I could not move a muscle. This, according to the book I had read, was the precursor to an out-of-the-body experience. So far so good, I thought, now for stage two. Then, by a sort of combination of imagination and will-power I attempted to float vertically upwards. The result was most interesting. The feeling was exactly as though I was embedded in the mud at the bottom of a river and the water was slowly reducing the viscosity. I gradually became free and began to float upwards. As though my eyes were now open, I observed the ceiling approach, I passed smoothly through it, noticing no obstruction, and entered the darkness of the space beneath the roof. I then floated freely through the tiles, my body still cataleptic and horizontal, and my speed gradually increased. I could see the moon and clouds quite clearly and to this day I remember the wind whistling through my hair as I shot rapidly up into the sky — horizontal, cataleptic, and dressed in pyjamas! (One writer, Yram, describes this experience and calls it 'skying'.) Arriving back in the body I wrote down all the details in my usual way.

Thinking over that experiment I decided that it was not of much value. Any sensible person would say that I had merely been dreaming. What to do? It was clearly necessary to attempt to go somewhere and obtain some evidence, information which I would not normally have. I could then return to the body, write it all down, and check the facts in the normal way during the following day.

So the second time (it took me two or three more nights to have success again) I stopped the movement when my body reached the height just below the ceiling and I changed the direction of willing and imagining so that I floated feet-first towards the foot of the bed. The aim was to float through the window frame and describe a parabola down on to the lawn. According to Muldoon the catalepsy should then disappear because the 'silver cord', which Muldoon said was rather like a hose pipe when the two bodies were close together, would be stretched out to a thin silvery line and the forces through it (whatever 'forces' might mean), would be reduced to a low magnitude so the catalepsy would disappear. One could then walk about — or float about, whatever was convenient — and determine the facts which were required.

This was a splendid scheme but it did not work out in practice. I floated happily through the window frame and was about to make the descent to the lawn when I felt two hands grab me over the ears and hold my head firmly between them. The two hands then moved me smoothly back into the room and pushed me downwards into the body. From that day to this I have been unable to determine any proper explanation for that experience. I can, however, suggest quite a number, ranging from discarnate entities watching over me to a dramatization of hands by George because there was some

'I felt two hands grab me over the ears'

unconscious reason why I should not complete the experiment. It remains a matter for conjecture. (Robert Monroe, whose work is referred to below, had several experiences of feeling guiding hands during his experiments in leaving the body.)

After spending an hour each night for a month carrying out those experiments I was becoming tired and inefficient at my daily work so I decided to discontinue the experiments for a while. From that day to this I do not seem again to have had the right combination of leisure, energy and time to continue. Perhaps one day!

Before turning to near-death experiences I must say one or two more things. The first is that considering the out-of-body experience as moving around the physical world in another subtler body which interpenetrates the normal physical body is a gross over-simplification of the facts. It is clearly not that for two reasons: first, one does not have the physical eyes and therefore presumably could not receive physical rays of light; and secondly, the pseudo-physical world observed often has changes which may be symbolic. As mentioned earlier, it is a sort of dramatized reconstruction of a memory of this world. However, the world as each of us knows it, is nowhere but in our minds: (this is discussed in detail in the final chapter). Our minds appear to be all linked together, as evidenced by telepathy. Thus it is clearly possible for someone who may be imagined as a mind temporarily disengaged from the physical body to share the 'physical world' of someone else — or at least a near-approximation to it. There is little doubt in my mind that the dramatizations of the out-of-body experience do appear at times to allow true information of normal physical facts to be paranormally acquired. Many psychics and others have the idea that the 'astral body' is made of subtle material interpenetrating the physical body and projecting all round (see 42 and 43). For the moment let us look at Robert Monroe's experiences in the out-of-body state and his suggested way of achieving it, together with a little about his work involving group exploration of that other level of consciousness.

ROBERT MONROE

Robert Monroe was in early middle age before he had his first and spontaneous experiences. Monroe is one of those rare people who try to study these unusual experiences objectively and he has worked with appropriate scientists in doing so. He started a research institute in the Blue Ridge Mountains in Virginia, where I visited him some years ago.

Monroe's early experiences were somewhat alarming. While lying down he appeared to be struck by a beam or ray out of the sky and this caused his body to shake violently, the body itself being cataleptic at the time. He had the greatest difficulty in forcing himself to sit up and only then did he slowly return to normal. He had this experience on a number of occasions and during one session the vibration seemed to develop into a ring of sparks sweeping to and fro around his body. As this ring passed over his head he experienced a roaring noise.On a later occasion, after the roaring noise and vibrations had occurred, he was waiting to go to sleep when he found that he could push his hand right through the rug and boards beside his bed and feel the ceiling of the room below. On another occasion Robert Monroe found himself floating in the air and gently bumping against the ceiling of the room, his body being visible in the bed below, his wife sleeping beside it. Doctors with whom he discussed these episodes were not very much help. At this stage Monroe, a business man in the United States and qualified in communication engineering, knew nothing of the fairly extensive literature on 'astral projection' in the West nor of the various 'occult' traditions concerning the same subject in the East.

Some of Monroe's experiments have been most intriguing. On one occasion he 'visited' in his other body a woman friend who was talking with two other younger women. He had a brief chat with her

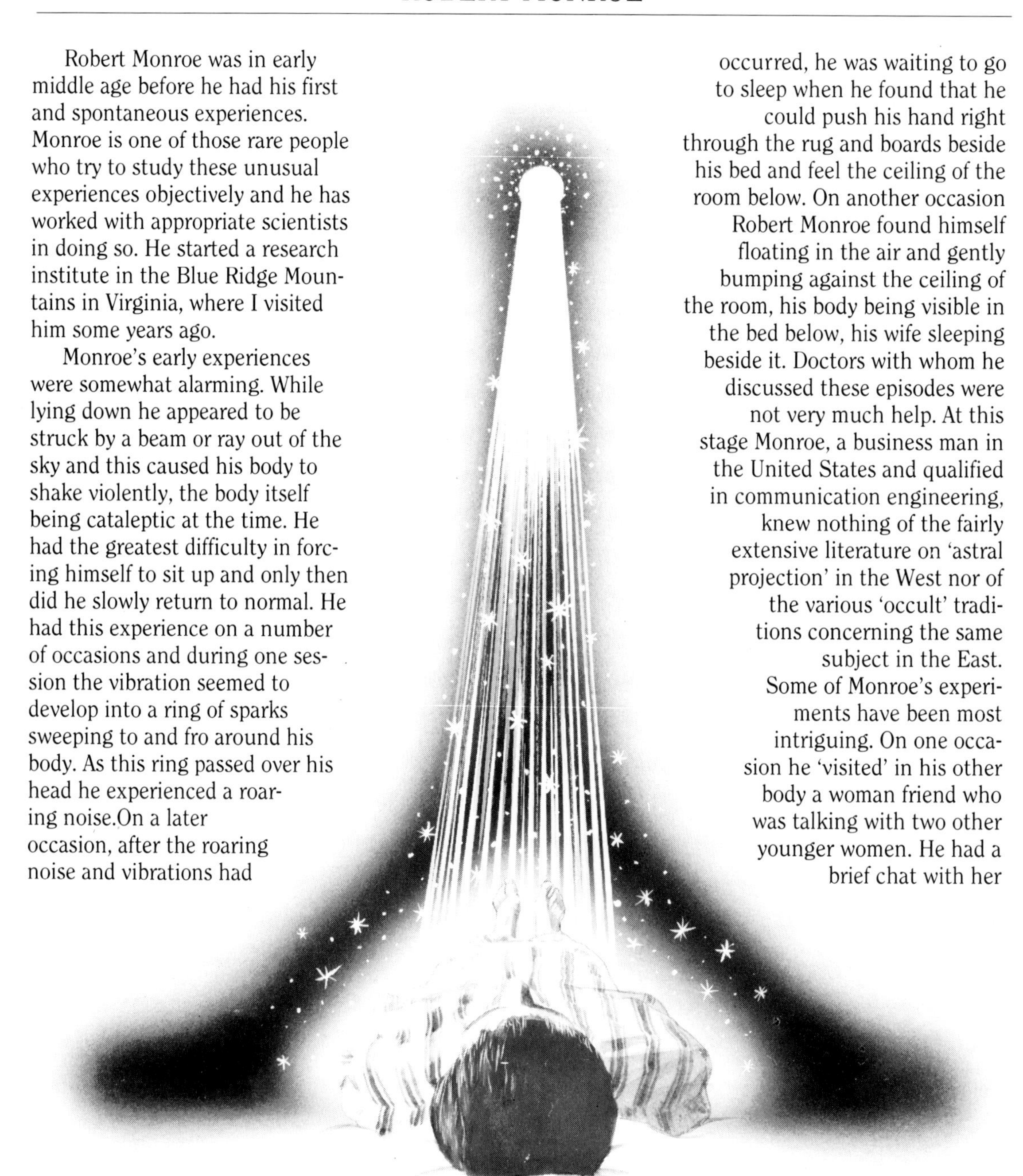

Robert Monroe experienced the sensation of being struck by a beam from the sky. A ring of sparks passed over his head, accompanied by a roaring noise.

while she continued to talk to the two others and then he said that he would make sure she would remember the visit by pinching her. He did this and she uttered a sharp cry. Three days later he checked with her and she reported that at the time he had been experimenting she was indeed with two other women doing exactly what he had 'observed' but she had no recollection of communicating with him. However, when he referred to the pinch she was greatly astonished and showed him two blue and white marks at about her waist level, which was the position where he had paranormally pinched her. She had no idea what had caused the pinch. Particularly intriguing in this story is the phenomenon, often experienced by Monroe, of having a 'conversation' at one level of someone's mind while the person concerned is fully conscious and is at the same time having a normal conversation with someone else. He or she then has no memory whatever of having paranormally communicated with Monroe in this way.

This shows very clearly that there can be more than one stream of consciousness proceeding at one time. It reminds me of an experiment I did many years ago in which I had a chat with an automatic-writing medium about normal things while her pencil was covering the paper with writing about something quite different, as from as ostensible communicator.

Monroe had many other most interesting and unusual experiences which are not, so far as I have noticed, mentioned in any other books on astral projection. For example, he once conducted an experiment within a Faraday cage (a room completely enclosed by wire mesh to prevent the ingress of radio interference) but in his case the cage was held at a steady potential above earth of fifty thousand volts. He attempted to project and found that though he did this successfully, leaving his physical body, he was unable to escape from the Faraday cage. He said that the feeling was as though he were held in a flexible wire bag, and he thinks that this was the result of the electric field. (He suggests further that this might be the basis of a 'ghost catcher'!) On another occasion, while projecting to a friend, Monroe found himself floating near the second floor level of a row of houses. When he went later (in the ordinary way) to inspect the site he observed that there were overhead cables carrying the electric supply to the houses along and near the path that he had found himself taking. He suggests that these might have produced some sort of attraction. (This all reminds me of the claim that some people seem paranormally to damage their electrical equipment.)

On one occasion Monroe tried to observe, (having heard about the 'silver cord') whether he had this well-known 'silver cord' joining the two bodies. He had not previously noticed this but found, when feeling behind him, that there was indeed a cord which came out from between the shoulder blades of his other body — to which it was rooted like the roots of a tree. The cord felt as though it consisted of a large number of parallel strands. When fairly close to his body it seemed to be about five centimetres (2 inches) in diameter.

Monroe often refers to the 'ideoplastic' nature of the second body and suggests that its arms can be stretched out to several times their length in order to feel something. Action and thought appear to go closely together.

He notices in his other body experiences that smell and taste appear to be absent and seeing and hearing are rather different: he can see in all directions at once and speaking to someone else is achieved by a form of telepathy — the words are made by the receiver of the thoughts. All this is of course not particularly noticed by someone who has the experience for the first time, just as all the finer points of normal perception are not necessarily observed. Though he mentions that any kind of body which is desired can be created (one can have the body of a dog or a bird), if there is no particular thought along these lines then the body tends to lapse back into its normal human shape.

A somewhat disturbing experience — which Monroe has had on more than one occasion — has been finding himself in the 'wrong body' on his return to physical-world consciousness. (On one occasion he was in hospital being helped to walk.) He immediately reprojects and thinks clearly of his own body: this is usually effective in returning him to the right place!

Monroe gives interesting and useful hints as to how one can train oneself to have an out-of-the-body experience (see *Journeys Out of the Body*).

Monroe identifies three kinds of space in which he finds himself as a result of various procedures. He refers to Locale I as the 'here, now' — the space just like the physical world. He refers to Locale II as an apparently infinitely large space in which are the traditional heavens and hells and which people appear to occupy after death. He refers to Locale III as the region in which people are living lives very much as they are on the earth only at what appears to be some quite different period of history. He describes a number of occasions when, in that third state of consciousness, he took over someone's body and found himself at rather a loss when trying to answer questions put to him by others. His remedy was rapidly to escape and allow the original occupant to return. It is hard to know what to make of such an experience and one is reminded of the Spiritualists' suggestions of possession here!

Monroe made a number of experiments in which he endeavoured to visit people he knew who had recently died and were in Locale II. He describes seeing them and their appearance being that of their physical lives at a younger age but he was not apparently able to have a conversation with them. He describes them as seeming to lead some sort of life rather like the one they had left.

Rather alarming fights with 'non-human entities' are described by Monroe. By use of appropriate thought imagery he manages to win the fights or appropriately escape, diving back into the physical body for safety. For instance, one piece of thought imagery which effectively rid him of a small pseudo-human body clinging to his back involved putting against it a pair of electrified wires. This deflated it like a balloon and he noticed a bat-like creature which flew away from him with a squeak.

When I visited Robert Monroe he showed me his laboratory in which he produces out-of-the-body experiences in other subjects by means of appropriate acoustic stimulation. I shall never forget the fascination of hearing about his recent work when he came to join me over breakfast at the Holiday Inn, in the mountains near his laboratory and to which I had flown the evening before from Washington, DC. He described how subjects are given out-of-body experiences as a result of lying on an air bed in a semi-darkened room, acoustic stimulation being applied through headphones. He told me — and it all sounded like science fiction — that members of his Experimental Group travel together into other regions of 'space', meeting and communicating with other humanoid characters — the communicator in one case being at a much higher level of intelligence than were the experimenters and treating them like little dogs which had wandered in off the street. It is very clear that, as Eastern traditional teachings suggest, in these altered states one is in a world of the mind where memory and imagination, conscious and unconscious, produce most confusing experiences, and with a certain content which does seem to be objective.

Robert Monroe was kind enough to give me a taste of his semi-sensory deprivation/acoustic conditioning and it did feel as though I might leave the body. He did not continue, however, explaining that the aim was merely to give me the flavour!

WAYS OF INDUCING THE OUT-OF-BODY EXPERIENCE

There are quite a number of ways of inducing the out-of-body experience and I propose very briefly to mention them. A deep state of relaxation is usually the necessary precursor, although I do know of exceptions to this. My own experiences were obtained as a result of deep physical relaxation combined with what I would describe as autohypnosis and the use of the imagination and will. An English researcher, Sue Blackmore, had her major experience at university as a result of being very tired and smoking hashish or 'hash' (sometimes referred to as cannabis, marijuana or 'grass'). No doubt other hallucinogens such as LSD might also produce the experience. It does not of course happen invariably and such experimentation with drugs would, in any case, be ill advised — and very dangerous.

Sensory deprivation is another way of having the experience and I well remember many years ago, when both the Americans and Soviets were carrying out experiments on sensory deprivation (in connection with preparations for space flight) that out-of-the-body experiences were often described. Those experiments involved the subject's being floated in water at blood temperature in a darkened and completely quiet environment, the aim being to eliminate input of any kind through any of the five senses. Trainee astronauts found themselves floating in 'another body', sometimes high in the air above the sensory-deprivation apparatus and from here they were able to observe what was going on at great distances and had other apparently psychic experiences.

It has been traditionally believed in India that if Raja Yoga (meditation) is practised correctly for an appropriate time then the so-called *siddhis* (that is Sanskrit for psychic powers) will appear. One of these psychic powers is described as the ability to leave the body at will. (The traditional book which explains all this is Patanjali's *Yoga Aphorisms.*) The aim of yoga (discussed further in chapter 8) is, as the word itself implies, to achieve union — union with the great source of all life behind and within the universe. The trainee yogi is warned against being deflected from the spiritual path by the coming of the psychic powers.

Much scientific work has been done to study one particular kind of meditation, the kind popularized by Maharishi Mahesh Yogi and known as transcendental meditation. Electroencephalographic (EEG) records have shown that this practice leads to what the TM (transcendental meditation) proponents call 'coherance' (their spelling) — a state of uniform brain rhythms all over the head, at front and back of both left and right hemispheres. This could be a precursor to an out-of-body experience but, so far as I know, experiments have not yet been carried out in that direction. I mention it because I understand that Robert Monroe's acoustic stimulation methods involve putting into each of the ears separately (via the headphones) a note of a slightly different frequency. The ears thus hear the notes separately but the brain combines them so that a 'beat note' is heard. This perhaps causes the same state of coherence which the TM people describe. Monroe adds to these notes various other sounds, notably the 'white noise' or rushing sound which he himself experienced during his own out-of-body experiences and if I remember accurately, he also adds a few bell-like sounds.

It should be noted that no matter what method is used, by no means everyone has the experience and the proportion for whom a particular method would be successful might be quite small. Statistics are not yet available as sufficiently wide research has not been carried out.

Meditation may be a precursor to out-of-the-body experience.

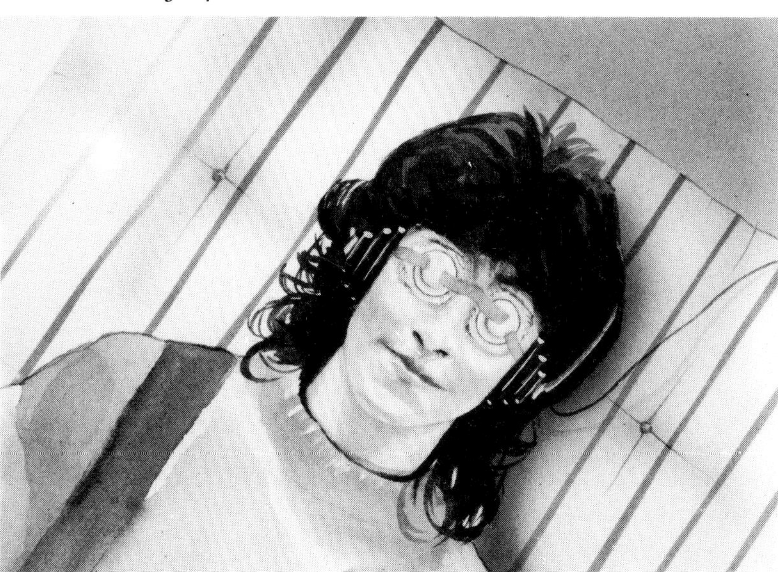

NEAR-DEATH EXPERIENCE

It has been possible during relatively recent years, with the growing volume of medical knowledge, for doctors to resuscitate a patient who is clinically dead. For many thousands of years, if someone died then they were pronounced dead and that was the end of the matter. Now a patient can have no heartbeat, no breathing, no electrical brain activity (as picked up by the electrodes of the electroencephalograph or EEG machine) and yet they can, if that state has not been in existence for too long, often be brought back to life. Usually the period during which there are no such vital signs of life is only a matter of minutes but there have been cases of much longer periods. I remember one occasion when a 'drowned' boy was trapped under the ice in a frozen lake for an hour or more. If the period

Advances in medicine mean that a clinically dead person may be revived and can sometimes describe vividly the near-death experience .

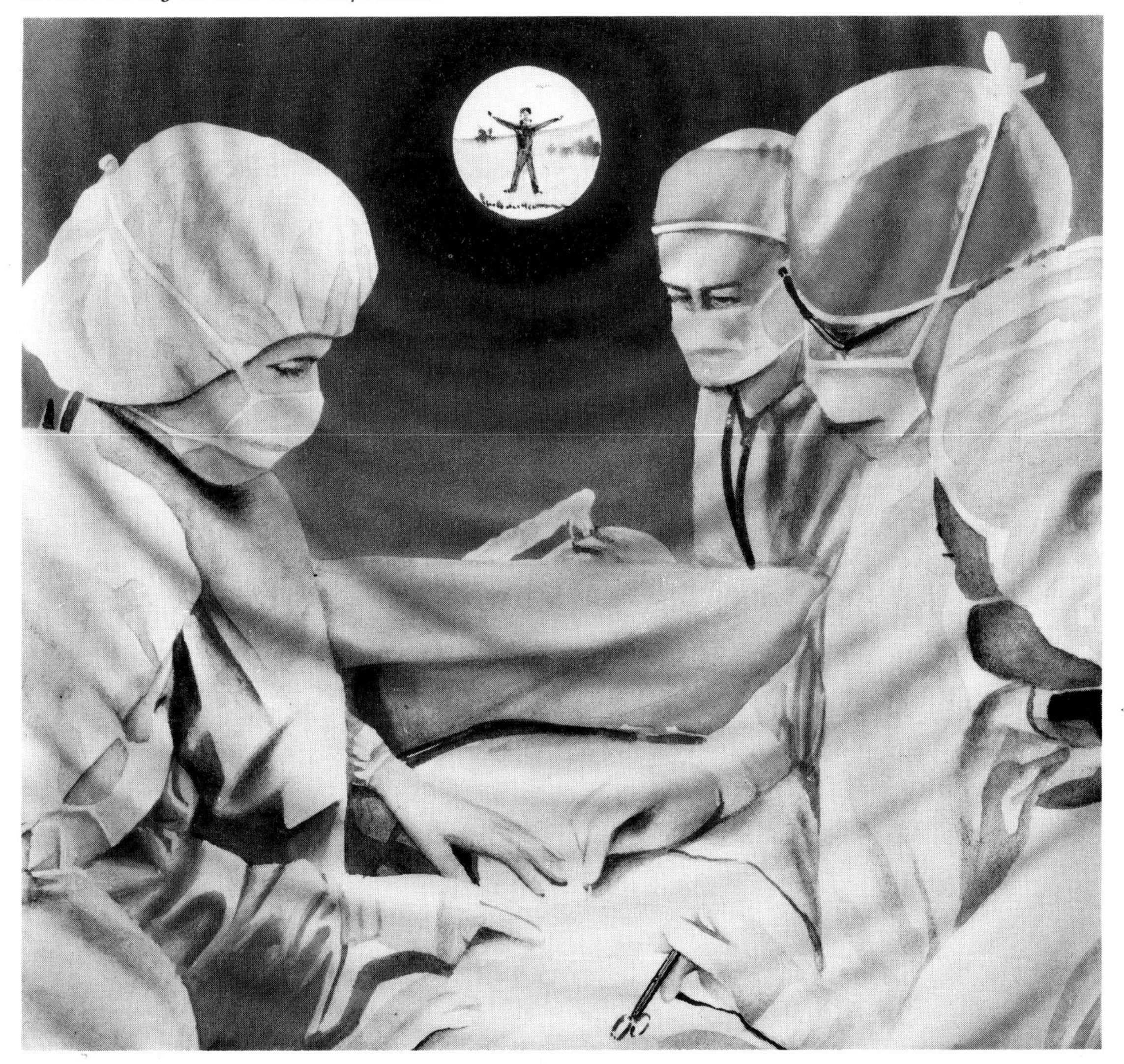

during which the heart is not beating is too long then brain damage will result from lack of oxygen. No doubt the case of the boy under the ice was an exception because of the very low temperature at which the boy's body was maintained.

The very materialistic Westerner would no doubt suspect that during a period of near-death a subject would have no experiences of any kind; that there would merely be a blank period in their lives. That is, however, by no means the case. In the last twenty years or so, several doctors have been asking patients who have been through the near-death state what experiences if any they had during that period. The result has been not at all what might have been expected by a naïve realist but more like that expected by a religious believer.

Sometimes, after hearing the doctor pronounce them dead the patients experience an uncomfortable buzzing noise and the impression of rapid movement down a dark tunnel. They come out into the light, seeing their own body from a distance and observing that they have another body. Then they see a kind of ball of light, what we might call a spirit, which appears before them. In the case of many people it is as described, a sort of ball of light. In the case of a believing Christian it might take the shape of Christ. A Hindu might see the glowing form of Sri Krishna. This 'spirit' communicates with them in a sort of telepathic way and runs over, with a pictorial review, the life which has just ended, evaluating it with them, especially the effects of their words and actions on other people. After this review, they move on and observe friends and relatives, already dead, coming to meet them. Sometimes they find themselves approaching a sort of symbolic boundary or border between this life and the next: it might be a gate in a wall, or a river. At about this point they may hear a voice telling them that they must return or perhaps that it is not yet time. Occasionally this appears to be a discarnate relative, perhaps a parent. Then they regain consciousness, finding that the doctors have made a final successful attempt to revive them.

People who have had that experience find great difficulty in putting it into words but usually their lives are drastically changed for the better as a result, with no doubt whatsoever about life continuing after death and a very clear appreciation of the importance of treating others in a kindly way and living so that the lives of others are made happier rather than the reverse.

I first read these facts in the book by Raymond Moody, Jr., in the summer of 1976 and soon afterwards I was invited to a dinner in London with a society of clergy. Finding myself seated next to the chaplain of a large London hospital I said to him, 'Do you sometimes talk to patients who have been clinically dead?' He replied, 'Yes, occasionally.' I then asked him whether any of those patients had described an experience anything like the pattern I have set out just above. I have rarely seen anyone look so astonished. He said, 'Yes, I certainly have. How on earth did you know that?' He thought that the particular patient he remembered had experienced delusions of a special kind and that it was probably unique. He was very surprised to find that it was by no means unique, that there were some hundreds of cases known at that time (and in the years since there must have been several thousand more which have been recorded).

Some of the experiences of near-death start with what appears to be a normal out-of-the-body experience, perhaps of floating in the air and observing the physical body down below being attended to after an accident, or during a surgical operation, or perhaps undergoing a cardiac arrest. Sometimes after that initial out-of-body experience there is the communication with the 'being of light' (as Raymond Moody calls that entity) and the review of the past life. The communication with the 'being of light' is always a joyful loving one and, although the activities being reviewed are sometimes indications of selfishness rather than benevolence, there is never any shadow of blame but rather the impression of more lessons learned. In one of Moody's cases the 'being of light' actually took the subject in the other body by the hand and travelled with him through the floors and walls of a hospital to show him what was later to occur, indicating that at some time in the future they would be there to meet and help him with the actual process of death.

I asked Robert Monroe whether any of his out-of-the-body Experimental Group had ever met the 'being of light'. He replied, 'Yes, often. There is no use becoming all reverent and bowing down otherwise the whole experience is at an end there and then. You have to go up to it, look it in the eye, shake it by the hand and go on from there.' (How one looks a ball of light in the eye and shakes it by the hand I cannot begin to guess, but it is no more mysterious than many other aspects of this altered state of consciousness.)

In the near-death experience, as in the out-of-body experience, thought is exceedingly clear. There is no muddle and confusion as there would be if the experience were the result of drugs. I mention this because many doctors following the different branches of their profession have suggested various interpretations of the experience, based on their own disciplines. However, so far as I can see, there is no simple and clear explanation based on ordinary physiological, biochemical or psychological lines.

By no means everyone who has a near-death experience discovers that they have a second body which is just like the first. Some people, for reasons which are not at all clear, find that they have no body, but appear to themselves like a cloud of mist.

Near-death experiences can be terrifying.

The predominant feeling for most people during a near-death experience is one of joy and delight and a strong desire not to have to go back. However, the sense of duty, perhaps towards a young family or a husband or wife who needs assistance, is a powerful incentive for returning. There is a definite sense in many cases of making a choice. It is not always another entity which says, 'You must go back!'

It is interesting to consider that when Raymond Moody was collecting details appearing in his first book *Life After Life,* Dr Elizabeth Kübler-Ross was doing the same. The facts they had each discovered (without knowledge of the other's work) agreed perfectly. A number of different books by doctors and other professional people have further confirmed the original facts and Michael Sabon's book in particular, *Recollections of Death,* is an enormously valuable collection of cases investigated by him as a cardiologist. (As might be expected, the cardiologists are the people who have most acquaintance with the near-death experience.) The major recollection I have from reading Michael Sabon's book is of the remarkably detailed descriptions given by the near-death experiencers he interviewed — in particular concerning the technical details of their own resuscitation which they could not possibly have unconsciously dramatized. The experience really does appear as if it provides, in its initial stages, a 'normal' view of the physical scenery and of the people and their activities in it.

While it is only during relatively recent years that we have all this scientific examination and assessment of the near-death experience, we do have ostensible descriptions of the death process recorded over many thousands of years. Plato's descriptions of the death process are very similar and include most of the elements which we have described above. Both the Egyptian and Tibetan *'Book of the Dead'* are in general agreement.

There is one other matter which must be mentioned. I have perhaps given the impression that every near-death experience is by and large very

pleasant. This is by no means always the case: occasionally there are some very unpleasant experiences reported. I remember one situation described by Margot Grey, when the subject found herself on a ledge in a gloomy cave, while down below were the traditional flames and brimstone of hell. On another ledge at the other side of the cave there was a tiger and she was very frightened that the tiger might spring across on to the ledge and dislodge her into the cauldron below.

Moreover, suicides do not find that they have escaped any of their problems but instead find that they are surrounded by the same anxieties but in a murky and miserable environment and with no means of correcting the situation which gave rise to their despair. It appears also that those who commit suicide through grief, in order to be with someone else who has died earlier, do not thereby join them but are separated probably for a longer period (at least it might *seem* longer) than if they had lived out their lives in the proper 'intended' way.

In all this, it is fairly clear that the experiences we are likely to have immediately after death are, in effect, a reflection of what we have made of ourselves. Our external surroundings appear to be a 'symbolic form' of an internal mental state, both conscious and unconscious. The religious stories of hell-fire, demons with pitchforks and the like may well be created for us by some unconscious mental process after death. And they would be perfectly 'real' to us in that state of consciousness.

Professor H.H. Price wrote a paper many years ago which appeared in the SPR *Proceedings* with his speculations on what the 'next world' would be like. His conclusions were based on the assumption that we still exist (the evidence being very good), that we can clearly have no input from any of the five senses but that we retain our memories and desires (conscious and unconscious). He arrived, along logical lines, at a 'next world' rather like that described by the Spiritualists and called by Monroe 'Locale II' (see page 78). (He went on to suggest that much the same logical difficulties would apply to our ordinary perceptions of the physical world.)

LUCID DREAMING – THE CHRISTOS EXPERIMENT

The Christos experiment is claimed to be a way of having 'conscious dreaming' or perhaps even of observing some of one's 'past incarnations'. I met the author of two books about this experiment, G.M. Glaskin, in London about ten years ago.

The Christos experiment is carried out as follows: The person to have the experience lies flat on his (or her) back on the floor with a cushion under the head and the shoes removed. Eyes are closed. Then the subject's ankles are massaged by another experimenter for several minutes to produce relaxation and soon after the ankle massage has begun another experimenter massages the area of the forehead between the eyes and above the nose, doing this very firmly with the outer edge of the closed fist. The massaging is carried out very vigorously. By using this prodedure, as well as taking a few deep breaths and performing a mental process of relaxation, the whole body is made exceedingly limp.

The next part of the procedure is similar to other methods which try to 'loosen' the normal attachment or identification of the mind with the body. Perhaps you remember the exercise suggested by Muldoon to loosen the attachment of the 'astral body' to the physical body by imagining, while lying on your back, that you are looking at the ceiling and then letting the gaze proceed down the wall across the floor, under the bed and up the other wall, the body meanwhile remaining still. In the Christos procedure the subject is asked to visualize his or her own legs and feet growing horizontally by about five centimetres (2 inches) to make the body longer. He then returns to his normal length and imagines his trunk and head doing the same thing, moving about five centimetres horizontally to make himself longer. Each of these exercises is repeated several times and then carried out again but for greater distances. Eventually he imagines his legs extending some forty centimetres (about 16 inches) and his head doing the same. He then imagines himself to expand all over, growing in size rather like a balloon. Assuming success with these exercises, the subject is then ready to 'project'.

The subject imagines himself – on the instructions of the experimenter who massaged his forehead – to be standing outside his front door, and describes it in detail. He is then asked to imagine himself standing on the roof of the house and is asked to describe it. Next he is asked to go straight up in the air and to describe the scenery from a height. All this visualization is continued by his being asked to imagine it is night-time and to describe the scenery then, afterwards changing back to day-time. Care is taken that the subject is clearly aware that he is doing the visualising, and is merely being guided by the other experimenter; that the experience can be changed or terminated at any time the subject chooses.

Following this preliminary set of exercises he is directed to descend to the ground and (if lucky!) will find himself in a quite different place and at a different time. He is then asked various questions about his feet, his face, his clothing, and the general surroundings. If all goes well, according to Glaskin the subject will have the most interesting experiences of finding that he (or she) is somebody else and will accrue information regarding either past incarnations or powers of dreaming; it must be very difficult to tell which. My guess would be that this is another example of George's dramatization as in normal dreaming. However, the experience is certainly very real and vivid for some subjects. Although it is carried out using a rather different procedure, it does remind one quite strongly of Muldoon's 'astral projection' and Monroe's 'travelling' (see pages 76 and 78).

Now I must describe my own experience as the subject of a Christos experiment. The technique used was exactly as I have described, with two other members of the SPR Council, who were close friends, carrying out the ankles and forehead massage. I was eventually asked to see whether I could 'project' to the house in Scotland of a former SPR president. (The experiment was being carried on in London.) I must confess that I had very little success in stretching or floating up in the air and observing the scenery around, although I did my best. When it came to the projection to Scotland I appeared to have no success at all but I did do my best to imagine it in the ordinary way. I informed the other experimenters of the thoughts which floated into my mind in an attempt to produce some sort of a result. One of the thoughts I had was of being in a room which had a table in the middle and a book on the table. The thought came to me that the book was *Travels with a Donkey* by R. L. Stevenson. After the experiment was over, one of the other experimenters telephoned Scotland to see whether any of this was true. Our friend in Scotland was quite surprised to hear this statement about the book. He replied, 'There is nothing on the table at the moment – but Stevenson's *Travels with a Donkey* was on that table yesterday.' No one could have been more astonished than I was!

Before concluding this piece I must say that similar rather astonishing occurrences have taken place during several experiments. I remember once being in the flat of a friend who had recently acquired a bust created by a sculptor acquaintance. Following a conversation about ESP I made an attempt to describe various aspects of the life of the person the bust depicted (whom I did not know), relating such things as political views, church affiliations, newspapers read, type of house, and so on. Nothing was written down but my friend in whose flat I carried out the experiment was quite surprised to tell me that much of what I had said was perfectly correct. This was hardly a scientific experiment but it did bring home to me that sometimes, perhaps more often than we do, we researchers should use ourselves as the guinea pigs.

Lucid dreaming while asleep

A sleeping lucid dream is another altered state of consciousness which has self-evident relationships to all that we have been discussing. Lucid dreams are dreams in which the subjects are well aware that they are dreaming and that the physical body is asleep in bed. There is normally in a lucid dream a complete memory of the subject's life up to that point and knowledge of the plans for the following day after waking up in the morning. A lucid dream is different from an out-of-body experience in that it is exceedingly rare for a lucid dream to involve an experience of looking at the dreamer's physical body from outside.

Lucid dreamers sometimes find that the environment can be consciously changed; they might find themselves to be standing in their garden at home and wish to change the scenery to say that of Piccadilly Circus. The procedure would be to close the eyes, strongly imagine that they are in Piccadilly Circus and open the eyes again. If successful, they will indeed find themselves in Piccadilly Circus and able to walk around (or float — whatever they choose) and generally examine the scenery.

It is possible to have various degrees of lucidity in a lucid dream and everything is not always perfectly clear; for example, memories and plans may sometimes be hazy. However, when a lucid dream really is fully lucid all sorts of possibilities open out.

It will be clear by now that lucid dreaming has a good deal in common with out-of-body experience or 'astral projection'.

Some laboratory study of lucid dreaming is being undertaken by physiologists and neurologists. It has been discovered that if a mild electric shock is applied to the forearm of someone who is fairly skilled in lucid dreaming it has the effect of waking them up in their dream to lucidity without actually waking them up to full consciousness in the normal way. When the dreamers have been thus awakened the experiments can then proceed in the following way.

During dreaming, as has already been mentioned, there are rapid movements of the eyes: the subject is in REM sleep. It is as though the activities in the dream are being followed with the ordinary eyes. It is found that the lucid dreamer can signal to the experimenter in the laboratory that he or she is lucid and is about to carry out the experiment. This can be done during the lucid dream by flicking the eyes a given number of times to left and right and this action will occur with the physical eyes as well as in the dream. The movement can be picked up by means of the electrical activity of the muscles of the eyes, these muscles producing the usual electrical activity which goes with their use. The eye flicking thus shows on an EMG (electromyograph) machine which records on paper the voltages appearing on the muscles when they are used.

A typical experiment of this type might involve the dreamer's lifting a weight and lowering it again, the activity of the physical muscles being observed by an EMG machine. The experiment would then involve the subject's signalling his or her lucidity and readiness to start the experiment. Then the experimenters would observe, for example, the signals from the calf muscles resulting. It has been found that a weight lifted in a lucid dream leads to the same kind of signals in the physical muscles as occur when the weight is lifted in the ordinary way, by someone who is conscious, but the signals will be of a reduced magnitude when dreaming.

This laboratory work on lucid dreaming has commenced only relatively recently and further results are awaited with interest.

HYPNAGOGIC AND HYPNOPOMPIC IMAGERY

Hypnagogic imagery occurs for many people just as they are on the way into sleep but before actually falling completely asleep; hypnopompic imagery can occur while they are waking up in the morning but before they are completely awake. A fair proportion of the population experience hypnagogic and hypnopompic imagery and a long paper about this subject appeared in the SPR *Proceedings* in the 1920s.

Some people in that deeply relaxed state midway between waking and sleeping find that they are observing (with closed eyes) panoramic scenes all around them. Sometimes they observe people apparently talking and see many different faces.

About one-third of people have these hypnagogic images on the way towards sleep. If they are in good health the images are usually happy, pleasant and very clear. In illness they sometimes become distorted and ugly or broken in pieces. People who are good at consciously visualizing a face or scene are no better than anyone else in regard to the appearance and clarity of hypnagogic imagery.

It is fairly common for the images to start with cloudy effects and then the clouds part, showing faces or scenes of the countryside. The faces are always vivid and living and in great detail: they are not just still pictures. In some cases the faces appear gradually bit by bit, perhaps the eyes first, followed by the teeth and nose and gradually the complete face. The eyes look at the percipient. Some people — I believe not a large proportion — can produce images of this kind at will. For example, Goethe could produce a rose — or rather a rosette — which was continuously putting forth petals from the centre. He could do this whenever he closed his eyes and wished it to happen. I know someone who can in this way produce objects, including working models, appearing to him totally real. He told me that he likes sometimes to go out of the room and see whether they are still there when he returns.

A most generally observed feature of the hypnagogic imagery is its endless and multitudinous variety. Landscapes are the commonest images for adults, followed probably by scenes with moving figures and inanimate objects. Landscapes are not seen at all by children. In cases of delirium and insanity the images are of animals, corpses, processions of the dead, devils and ghosts.

Sometimes the scenery is changing rapidly and is as if observed from a train. Often the images tend to take a circular form and very rarely they are seen as though from the back of the head. They are always bathed in the brightest and clearest light; they are occasionally in black and white but more usually in bright colours and with a rather strange luminosity. It is interesting to note that they seem not to be memories: many hypnagogic imagers have said that they have no memory of ever having seen the images before.

In the case of one imager he describes a number of young women passing in his images and said that they looked towards him and passed on. Then one of them spoke and the voice, clear, soft and distinct said, 'He isn't asleep'. Hearing is rather rarer than seeing in these images, but smell is even rarer and sensation more so. Music is rarer than voices.

Where the images come from is not at all clear and there are various theories. Some people think that they are scenes from earlier incarnations. Occasionally the images seem to be telepathic and some premonitory. Approximately fifteen per cent of people can perceive (but not necessarily every time) images in a crystal while in apparently normal waking consciousness. Such crystal images are very similar to hypnagogic images.

Hypnopompic images are images seen while in the process of waking up and are probably enduring memories of a dream that is going on immediately before awakening. They are usually not so interesting as hypnagogic images.

Using the iceberg model of the mind (see page 11) we would say that perhaps George is producing the images. What is the reasoning behind this? Perhaps it is just a game! Certainly the people who look at the images often take a great delight in them. The Ganzfeld experience described on page 146 has a certain similarity to hypnagogic imagery.

Regarding my own experiments in this area I had the following experiences. Some years ago, as a result of practising meditation for a longish period, when lying in a comfortable armchair with my feet up in a semi-darkened room I was able to produce a state of deep relaxation in which thought ceased almost completely: clearly there must have been a minute proportion of my mind rejecting stray thoughts which tended to float in. The result of this state of extreme tranquillity was interesting. Suddenly rectangular pictures appeared in front of me as though projected on to a screen. In my case the pictures were not moving but they did seem to be symbolic, as though George was trying to convey some sort of message to me from a deeper mental level. I had a series of images over a considerable period of time, starting during the deep relaxation process I have described but continuing in the form of imagery which appeared while I was lying in bed (after having slept) and, in one or two cases, when I was sitting relaxed with eyes closed on a train. The first image I remember was of a brick wall with bright sunshine on it, the brick wall stretching right across my field of vision. One does not have to be a psychologist to understand the symbolism of that! Some time later I saw an image in the middle of the night of a slipper which was embroidered in silver and gold thread, with other brilliant coloured embroidery. It was an eastern-style slipper with the toe pointed and turned up. A later symbol which I observed while relaxing in an armchair was of a table covered with a snowy white table-cloth, set for tea. There was a view of some delightful countryside through a window above the table, the view being partly obscured by snowy white lace curtains. A final image in the series, which appeared in the middle of the night, was of a beautiful bowl of red roses, each rose having on it tiny drops of glistening dew. The flowers were so real and vivid that it seemed as though I could reach out a hand and touch them. I never remember seeing anything so beautiful.

An interesting feature of all these images was that they remained visible only so long as I gazed at them without thought. The moment I attempted to reason about them or describe them to myself with the aim of remembering more clearly, they flicked off instantly.

CONCLUSIONS

From the experiences considered in this chapter you will see the remarkably creative propensities of the mind — especially when freed from its routine duties of processing data concerned with normal life. The great importance of symbolism is evident as expressing, in terms of physical-world objects and activities, deeper ideas and 'guidance'. Symbolism and myth are of course the basis of both religion and poetry. There does seem to be evidence that there are deeper parts of ourselves (or perhaps it is we who are the 'parts') which at times, if we allow them, will give us advice and guidance (and pleasure and joy).

The experiences of the near-dead and resuscitated appear to show us what it is like to die. The sceptic who suggests that those subjects were not really dead certainly has a valid point; for myself, I cannot see why there should be a radical change to oblivion if the already clinically dead body is merely allowed to become colder and to decay rather than being resuscitated. And the experiences agree remarkably well with many world-wide religious and other traditions concerning death and after.

The material of this chapter also provides food for thought on the nature and reality of the physical world, including our physical bodies. One is, it seems to me, compelled by the scientific evidence to consider whether the teachings of Hinduism (and of some other religions) which inform us that the physical world is in fact a 'maya' or illusion may not be, in an important sense, true. (It is suggested in the final chapter that the physical world of our experience here could be considered as a sort of teaching machine.) Especially important to us may be the indications that the life following this one (if you accept the evidence that there are in fact further levels of consciousness) is to some considerable degree the result of what we have made of ourselves in this life — reaping as we have sown. It may be a good idea, before it is too late, to give careful thought to the sort of life we are leading now! Those who have been through the near-death experience including the review and assessment of their lives with the 'being of light' do seem to be 'better' people as a result; kinder, more helpful and less selfish. It is interesting that this sort of life is also advocated by great religious teachers. The Sermon on the Mount of Christianity and the Noble Eightfold Path of Buddhism provide examples. The 'deconditioning' experienced by the astronauts observing the earth from space seem to me a similar effect.

Reality depends on one's viewpoint. We 'know' the world to be larger than an apple, their similar size here is an illusion: the picture is just one truth. Many astronauts have radically changed their views on life.

7
WHAT IS IT LIKE TO BE PSYCHIC?

Over many years I have known numerous psychics of all kinds, have had many conversations with them about their experiences and many of them have been kind enough to allow me to undertake experiments with them.

The first thing I have noticed is that the most highly gifted psychics appear to have been psychic from birth. It is possible to develop psychic powers (and this development is discussed in chapter 8) but it is rather rare for an outstandingly good psychic to be produced in this way.

When they are small, children who are psychic usually do not realize that their perceptions are any different from those of anyone else. One gifted psychic acquaintance, Phoebe Payne, who was married to a London psychiatrist, Dr L.J. Bendit, told me that as a small child she thought adults were very strange and inconsistent in that when a visitor arrived at the front door and knocked or rang the bell they were invited in and politely asked to sit down whereas, she said, when they floated in through the walls they were just ignored. In fact, she said, the adults were so very rude sometimes as to sit in the same chair with them. It was clearly not long before she realized that her perceptions were a little different from the average! A later experience she had was of 'talking to her grandma' while the coffin containing her body was being carried out through the front door watched by the distressed relatives. She excitedly shouted down to them through the window, 'Don't be so sad: Grandma is here with me.'

Phoebe's reaction to her grandmother's death.

Some years ago I was sitting in my room at my first university, preparing a lecture, when there came a tap on the door. It was a Wednesday lunchtime and I was not expecting anyone in particular. I called out, 'Come in!', the door opened and a young lady rather tentatively looked round it. Her name was Margaret and she had started work here the previous Monday (as a secretary in the departmental office). I had passed the time of day with her perhaps only once. Having no idea at all what she wanted I said, 'Do come in! Please sit down! What can I do for you?' She explained, 'I'm psychic and I hear voices. I have a problem and my voices told me that if I went out of the office, turned to the right, walked to the top of the passage and tapped on the door at the end I would find someone who could help me with my problem.'

Margaret explained that she not infrequently received strong intimations of disaster to befall various friends of hers and she did not know whether she should warn them or not. At that moment she had a strong presentiment that her fiancé was going to have an accident with the car.

I suggested that she should on no account warn people when she had presentiments of disaster. Warning someone might put the person concerned into a state of fear and trepidation and could actually cause the disaster she feared. I explained further that psychic information was just about the most unreliable form of information available; one never knew whether it was correct until verified. Much better to keep her own council and see what happened. (I must add that some psychics would disagree with my advice.)

So she went away greatly reassured and told me a couple of days later that her fiancé's car had suffered a small scrape in traffic but no real harm had been done.

That story shows the sort of situation in which psychic people often find themselves. They are subject to all sorts of feelings and presentiments about other people and it is sometimes quite a problem to know what to do.

'Being psychic' means a great many things and ranges over a very wide spectrum, from the more or less common 'hunches' which many people have to full-scale trance and materializations. The very best and most powerful psychics have usually been psychic from birth and may not at first appreciate that they are in any way different from other people. Like the psychic who could not understand the inconsistency of grown-ups' response to visitors (see page 88), most psychics have similar stories to recount. Many of them have had experiences of talking to and observing someone who had recently 'died' and had some difficulty in understanding why the grown-ups were so sad. Some psychics had apparitional playmates who came to them when they were children and played with them, the adults telling the children not to make up tales when they described these playmates.

Perhaps many more children than we suspect start off by being psychic and then, as they become adapted to the normal life of the physical world and are firmly conditioned by their parents to believe that psychic functioning does not occur, the psychism disappears. One psychic researcher of my acquaintance, Ernesto Spinelli, a child psychologist, did experiments on telepathy with children (for which he was awarded a Ph.D. degree) and discovered that it was strongest between very young children and became weaker as they grew older. (This work is mentioned earlier.)

It may be that the brain of a strong psychic is a little different from the brain of the average person. Aldous Huxley looked upon the brain, with the five senses, as a sort of reducing valve or series of 'band-pass filters', filtering out from the 'universe at large' the physical world. Certainly there appear to be physiological correlates of all mental states. A friend of mine who is a neuropsychiatrist, Dr P. Fenwick, measured the EEG activity (the electrical brain rhythms) of a number of psychics and found certain anomalous behaviour in the right temporal region of the brain in a high proportion of them. In fact some of those psychics actually remembered falling on their heads. One well-known American psychic became suddenly psychic after falling from a ladder on to his head. The mother of a well-known British psychic had an electric shock before he was born. (This is hardly scientific proof but it may be significant as it does appear to form part of a pattern.)

THE PSYCHIC'S VISION

Many psychics find, some from their earliest years, that they can see 'auras', that is, coloured nimbuses surrounding other people. The famous psychic Eileen Garrett, called such auras 'surrounds'. Most psychics observe 'other people' not seen by most of us and hear voices. Often they learn to keep rather quiet about their voices and visions as much of the medical profession is deplorably ignorant about such matters and a psychic would either be prescribed tranquillizers or directed towards a psychiatrist (who would probably know as little about psychism as the GP who prescribed the tranquillizers).

Most of the psychic people I know — and I have known a very large number over the years — have all been thoroughly normal, healthy and happy people. They just happen to have this extra window on to the universe.

Eileen Garrett used to see as a child a 'little old lady' who never spoke but who 'visited' her chest of drawers. Eileen Garrett could shut out the voice of her sharp-tongued aunt so that she could see her lips moving but hear nothing. She could also feel a 'oneness' with all light and life. She had three 'spirit' children who came to play with her as a child — two little girls and a boy — and they came until Eileen Garrett was thirteen. Everything and everybody grew up but not those children.

Eileen Garrett could see ordinary human beings surrounded by the nimbus of light mentioned above, but her children seemed to consist entirely of this light. They were 'soft and warm' when she touched them and appeared and disappeared quite suddenly. They strongly disapproved of her climbing trees and felt that she should leave them intact. Eileen Garrett describes a number of examples of creatures dying — both animal and human — and observing a sort of smoky substance float out of them at about the time of death to assume their shape before disappearing.

A number of psychics I have known, including Eileen Garrett, have claimed to be able to observe the coloured nimbus or astral body (some refer to an 'etheric body' synonymously and some as though it were a separate thing altogether) and stated that they could tell the mood of a person by observing the colours in the astral body. Disease was indicated by grey patches. I have known two psychics who described to me their work with doctors: from their psychic observation of the 'etheric double' they have been able to indicate where the doctors should X-ray or otherwise examine for disease. Unfortunately, as is so often the case with psychics, the claimed work never appeared in the form of refereed papers in reputable journals, so far as I know, and has disappeared without trace. To be fair, I can imagine what the editor of an orthodox medical journal would say if he received an article on medical diagnosis written in collaboration with a psychic! So perhaps the doctors concerned might be forgiven. One can always publish such material — if it is well and scientifically done and the author is ready for objective refereeing by experts — in one of the well-known parapsychological journals. (Unfortunately, when it is written by doctors, such material may have good medical context but is often full of rather naïve parapsychology!)

The mediums sat together, a screen between them, and wrote down what they could 'see'.

A great many psychics who have written autobiographies describe their early puzzlement and how eventually they found their way to a Spiritualist church where 'all was revealed'. The result of this is that usually the psychics become Spiritualists, with all that entails of both philosophy (which is excellent) and theory (which is not — it simply does not explain all the phenomena). I have discussed in an earlier chapter (on pages 42-43) the model which most psychics have of what occurs during a séance when they are using clairvoyance and clairaudience; the discarnate communicators are described as coming to the medium's room in their astral bodies and communicating by hallucinatory visions and voices. It all seems very simple and obvious until the discrepancies are pointed out.

I have many times been sitting in séances surrounded by several psychics and have asked them to describe for me what they are seeing in the room. Sometimes there are similarities but at other times their perceptions are rather different. (Mrs Brown will perhaps explain to me quietly afterwards that she is of course working on a 'much higher vibration' than is Mrs Smith!)

I once spent an evening doing a careful experiment with two mediums side by side to study this particular point of psychic objectivity. I arranged the mediums in my room at the university in two armchairs side by side but with a screen between them. I equipped each with pencil and paper and asked them to describe as accurately as they could in writing any people they could see in front of them whom I could not. They did this and, though I found certain rough similarities in their descriptions, they differed quite widely. Being psychic is by no means observing with clairvoyance an objective astral world laid out all around. This statement of course agrees with other well-authenticated facts. I remember one lady psychic observing my reading a well-known book containing coloured paintings of the auras of people in various degrees of spiritual development and mood. She said 'I don't see them at all like that.'

Some psychics have visions of more or less traditional fairies complete with gossamer wings and wands. They sometimes suggest that as the bodies of the fairies are made of 'etheric material' they can assume any shape they like at will. Sometimes they are observed as points of light flitting to and fro amongst the flowers and other growing things and it has traditionally been believed that fairies are a low form of life, something to do with the growth of plants. I do assure the reader that otherwise normal individuals do sometimes have experiences of this kind. The stories of fairies, gnomes, elves, undines (sprites), and other like creatures presumably had some sort of origin and it is of great interest to know that psychics sometimes perceive the traditional forms. This is not something to be dismissed but surely something yet to be understood. We are clearly again, as so often, in the world of the mind and the mind is exceedingly creative.

A well-known psychic, Geoffrey Hodson, with whom I at one time conducted some experiments, described enormous angelic creatures he observed and with whom he communicated. He was a sensible well-educated man who enjoyed any opportunity to work with scientists.

Lilian Bailey, who was one of the best-known British mediums, makes an interesting point with regard to her development of mediumship. She gradually became involved with several mediums and began to have experiences: she remarks that if the 'buzzing' she heard engulfed her she would be in trance (and she was at first fighting against it). This seems to be a similar sound to that mentioned earlier in connection both with the near-death and the out-of-body experience.

Lilian's description, later on in her career, of 'giving clairvoyance' in a public meeting is interesting. She explained that the 'spirit people' she observed near the members of the audience (and assumed to be their deceased friends and relatives) were clothed only if the clothing were useful for giving information — such as a service uniform — otherwise she normally observed head and shoulders emerging from what looked like a sort of 'ectoplasmic draping'. She says that she could not see the figures when her eyes were closed and therefore assumed that they were objective. Sometimes they appeared to move instantaneously from the back of the hall to the platform beside her. Some 'spirits' found communication difficult, others easy, depending on their personality and the conditions in which they normally lived. She says that it helped considerably if the recipient for whom she assumed

At a public clairvoyance meeting Lilian Bailey sometimes saw spirit forms associated with the audience.

they were endeavouring to communicate spoke up loudly and clearly – otherwise the spirits tended to fade away and become inaudible to her.

Some psychics describe rather remarkably different phenomena from those of others. Eileen Garrett, for example, saw at an early age 'globules of light bursting within beams of sunlight' and she noticed that these globules appeared to be absorbed by the 'surrounds' of living things – animals and plants. The traditional picture is of *'prana'* (well-known in the East as an energy-giving subtle 'force') as vivifying the 'etheric doubles' of all living creatures. Psychic healers sometimes consider that they are renewing and refreshing the 'charge' of *prana* in the etheric bodies of their patients. (Prana is of course not known to Western science – see also page 119 .)

Eileen Garratt was also able to project what she describes as a 'flowing part of herself' into the personalities of people and things and even into distant countries. She noticed something else which is not too often mentioned – that conversations and music produced to her vision flowing colours in space. Besant and Leadbeater also described this phenomenon.

Annie Besant and C.W. Leadbeater (two well-known and distinguished early members of The Theosophical Society) developed psychic faculties later on in life, after meeting Madame Blavatsky, and described exceedingly unusual experiences. Besant and Leadbeater had experiences of 'travelling to other planets' and 'observing' the life on them. Needless to say, all this does not appear to make sense from a scientific point of view, in particular in the light of later information obtained by space probes. It would appear to be the results of the dramatizing capabilities of 'George'. (Madame Blavatsky herself warned of possible self-deception resulting from this: she used a Sanskrit term for the dramatizing capabilities of the unconscious: 'unconscious *kriyashakti*'.)

Besant and Leadbeater did research together using psychic examination of chemical atoms and employing what they described as 'the magnifying capability of the *ajna chakra*'. By this they meant the Eastern theory that the 'subtle body' has within it seven vortices called *chakras*. It is the vivification of these by the 'serpent fire, *kundalini*' (another 'force' referred to in Eastern literature) that leads to the development of psychic powers.

THE COTTINGLEY FAIRIES

These photographs, taken in the spring of 1917 and published in the Christmas edition of *Strand Magazine* resulted in a storm of controversy.

The two young girls, cousins then aged ten and thirteen, now admit that they faked the photographic plates but their skills were sufficient to baffle the experts of the day and their methods eluded controls placed on the production of a further set of photographs.

Perhaps the most interesting fact is that the appearance of the fairies in the Press evoked a flood of letters from other people who claimed to have seen similar figures.

After Besant and Leadbeater had developed their psychism they were apparently able to perceive the chakras referred to earlier. Probably their own development method involved the traditional 'vivification of these chakras by *kundalini*', assisted by their Gurus. The details are by no means clear in the literature!

The roaring and buzzing noise referred to so often would perhaps be described by an expert in *kundalini* as the subjective results of raising this psychic force and having it directed around the chakras, leading to the psychic capabilities.

Following this particular theory of *kundalini* and using Eastern methods of self-development undoubtedly led to remarkable psychic experiences — whether or not it 'makes sense' along Western scientific lines.

PSYCHOMETRY, 'MASTERS' AND ASTRAL STRUCTURES

To complete this discussion of what it is like to be psychic we must describe one or two matters not already mentioned. The Dutch psychic (or paragnost, as he called himself), Gerard Croiset, claimed that he helped police forces all over the world to find missing persons. (Other psychics have apparently done the same.) He would be telephoned in his home in Utrecht and would straightaway begin to receive images. His descriptions were, he told me, often of help to the police in guiding their search. If he had a personal belonging of the missing person this would help the images to appear.

This latter faculty is called 'psychometry' and is frequently demonstrated in Spiritualist organizations (but not always with great success). It was studied by Hettinger using an interesting statistical evaluation to assess what success might be expected by chance. (A London University Ph.D. degree was awarded for this work.) His results using two psychics were claimed by him to be statistically significant. Some of the information he obtained in this way concerned the experiences of the owner after he had parted from the article. The French researcher, Dr Osty, obtained similar results.

Annie Besant and Charles Leadbeater also described leaving their physical bodies during sleep and going astrally to a certain valley in Tibet where they were taught by 'Masters', that is, beings at a much more elevated spiritual level than they were themselves. Certain distinguished fellow psychics and others considered this to be another example of unconscious *kriyashakti* or, as we might put it, George's unconscious dramatization. I remember a fine furore about this! There is no doubt about the elevated nature of the 'teachings' obtained in this way — at least in my view — whatever validity one attaches to the descriptions of how the teachings were obtained. (Psychologists will perhaps be tempted to equate the Theosophical Masters with that 'archetype of the collective unconscious' referred to by Jung as the 'wise old man'.)

Some psychics can observe coloured forms being produced when someone speaks or sings — or plays an instrument (see also page 92). Leadbeater describes astral structures of great beauty produced by a symphony. He describes similar structures over a church produced by the singing — and especially by the sacraments. He further describes 'great Angels' present during the sacraments. One hardly knows what to make of this. If accurate it would greatly improve the interest of many scientific people, and others, in the ceremonial of the church. Rather more fundamental studies will need to be made in future!

An important idea that seemed to me to make sense of the way the psychic sees the astral bodies of people who are physically conscious as coloured ovoids was given to me by Michael Whiteman, the scientist/mystic. He tells me that in his experience any transfer of the centre of consciousness from one level (say the physical) to another (say the astral) immediately produces, almost automatically, an appropriate body for functioning at that level of consciousness — whereas normally there is not a body at that level, but merely a kind of centre or nucleus. Clearly most of us cannot make comments on such a statement from personal experience — at least not yet — but I know Michael Whiteman to be a distinguished and honest man of considerable integrity and with firsthand experience denied to most of us.

PSYCHIC DISCOVERY

When the well-known British psychic, Matthew Manning, was young many physical phenomena occurred around him, including the movement of many small objects which changed their place in the family home. These poltergeist phenomena occurred around him from about the age of eleven and became more distinctive, with larger objects being moved, as he grew older. The phenomena continued at his boarding school when he was fifteen. Objects of ever-increasing size, including cupboards and beds, were moved — sometimes several metres across a room. Scribbles and other writing appeared on the walls in the Manning home. Later Matthew put himself into a sort of trance and heard voices claiming to be those of people who had died a few hundred years ago. He then found himself apparently standing in the hall of the family house, observing former occupants, the house clearly having been modified in the interim period. In time he discovered how to do automatic writing and received 'communications' in this way. He discovered that practising automatic writing tended to stop the other phenomena occurring.

Matthew also described how he could see auras by switching himself into the state he found he required for automatic writing and he discovered particular colours to be related to certain facets of character, with dark patches indicating disease (agreeing thus with other psychics). Matthew found later that he could produce remarkable automatic drawings very like those of deceased artists, especially Albrecht Dürer. Some of his drawings were of existing pictures which he had never seen. Matthew Manning worked out for himself that he was in some way responsible for all these happenings though certainly not consciously. He also received automatically-written letters addressed to people he did not know, referring to people they knew and he did not, and ostensibly communicated by someone dead who knew them all! All this illustrates well how psychic phenomena occurred in the case of one quite remarkable and intelligent British psychic and how the development and bringing of the phenomena under control took place. Matthew Manning is at present running a healing centre.

Uri Geller has been mentioned earlier. He discovered psychic occurrences happening around him when he was very young. He describes how spoons and other cutlery used to bend spontaneously on the meal table, and his watch behaved erratically. Also, he found he had an internal screen (rather like the one I discovered in myself — see page 86) and this screen would, he claims, often show him drawings which had been presented to him in sealed envelopes. All these phenomena he used as part of his stage show later and observed that large numbers of other people could bend spoons too, after his encouragement, by wishing and believing that they would bend. One psychic occurrence Geller remembers well occurred when he was a small boy; it was of seeing a brilliant light in the sky. He considered that this event had something to do with his psychic propensities.

The important fact concerning Geller's metal bending — and as discussed on page 58, I am quite sure he could occasionally bend metal paranormally — is that he has no idea at all how it is done. He merely wishes and believes, and it happens. And he is certainly not deceiving himself: several distinguished scientists have carried out successful properly controlled experiments with him. One can safely ignore the magicians who say that it is all done by magic and that they can do anything he can do. This claim is quite simply not true and about as accurate as many of the other claims they sometimes make about psychics. (This rather sweeping statement certainly does not apply to all magicians: some freely admit that although they are able to imitate some of the Geller phenomena under their own conditions they certainly cannot reproduce all of them under the conditions often accepted by Geller in scientific laboratories.)

C.G. JUNG

No discussion of what it is like to be psychic could be complete without a reference to the great psychologist C.G. Jung. Jung's interpretation of his psychic experiences are, needless to say, quite different from those of most, if not all, others. Jung first dreamed of a 'wise old man archetype' whom he called Philemon. He considered him to be an ancient wise part of himself. Later he used to walk up and down his lawn having discussions with Philemon and he learned a great deal as a result. One is reminded of Socrates' daemon, who made Socrates one of the wisest men in Greece.

Jung says, 'Philemon and other figures of my fantasies brought home to me the crucial insight that there are things in the psyche which I do not produce but which produce themselves and have their own life. Philemon represented a force which was not myself.... Psychologically, Philemon represented superior insight. He was a mysterious figure to me. At times he seemed quite real, as if he were a living personality. I was walking up and down the garden with him and to me he was what the Indians call a Guru.' He refers to a highly cultivated elderly Indian, a friend of Ghandi's, who visited him fifteen years later, with whom he discussed the relationship between Guru and *chela* (disciple). The Indian said, to Jung's amazement, that his own Guru was Shankaracharya (who had died centuries earlier). 'There are ghostly Gurus too', he added. 'Most people have living Gurus. But there are always some who have a spirit for a teacher.' One is here surely reminded of the Theosophical Mahatmas, the teachers of Madame Blavatsky, Besant, and Leadbeater.

During an argument with Freud about the occult, Jung found he was able to produce a loud bang from a nearby bookcase. This was the apparent result of a 'curious sensation' he had which felt as though his diaphragm was made of iron and becoming red hot. He cited the bang to Freud as an example of 'so-called catalytic exteriorisation phenomena'. Freud said that was 'sheer bosh', whereat Jung predicted, not knowing what gave him the certainty, the immediate production of another loud report. It occurred — to Freud's astonishment — and the incident was never afterwards discussed!

THEORETICAL EXPLANATIONS

As mentioned earlier, psychics are not scientists: they assume things to be rather as they appear, just as the rest of us do in 'ordinary' consciousness. Psychics generally assume that if they 'see' a person and that person informs them that they are dead, then they are. And if that person is their recently deceased grandmother, then perhaps it may be so. (The evidence for survival is exceedingly good.)

However, the 'communications' from Soal's imaginary character John Ferguson (see page 44) show that it is not always possible (if ever!) to tell the difference between a 'thought form' and a 'real astral body of a discarnate person'. In fact sometimes communications are received from living people who are doing something quite different at the time. There is a well-known case (the Gordon Davis case) in which Soal received information from an acquaintance who said he was dead but who was actually sitting in his estate agent's office in Brighton at the time, quite unaware of anything unusual. Soal described receiving detailed information about a house in which the 'discarnate' Gordon Davis said he had lived. Three years later, after Soal discovered that his rather vague information about Gordon Davis's death in the war was not true, he found that at the time of the séance the house described had been chosen by the living Gordon Davis but no thought at all had been given to its internal arrangements. Yet the 'communicator' described not only the exterior features of the house but also the type of pictures and the placing and arrangement of ornaments. So a proportion of the information turned out to be precognitive. Information is sometimes obtained in séances about living people — but not usually in the first person!

It must not be forgotten, of course, that some excellent evidence for survival has been obtained via a number of mediums having psychic experiences so there is no doubt of their value. The point is merely that things are not quite so simple as psychics sometimes appear to think.

During platform or private clairvoyance communicators are said by the psychics to 'show themselves'. Lilian Bailey, as mentioned earlier, used to see the communicators as head and shoulders in a sort of cloud of whitish substance. How could she possibly tell that this was a genuine communicator and not a way adopted by her George to make overt information which had been obtained telepathically from the sitter?

Mediums, it should be mentioned, have discussed these matters with me on many occasions and often they have said that what I suggest is probably true for many mediums but not in their case, as being trained psychics they can tell the difference between a thought form and a genuine communicator. What do you think?

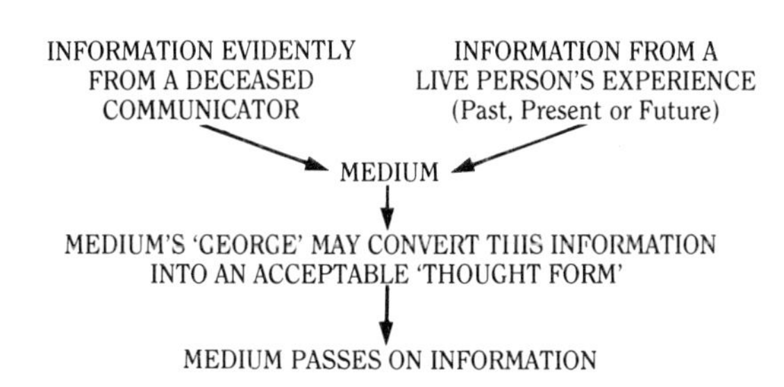

'CONTROLS' AND ASTRAL BODIES

Another claim which is often made by trance psychics is that the psychic is somehow removed right out of the way when his or her body is being used by a 'control'. This control lives in the next world, can see the communicators and has no difficulty in passing on information from them accurately. The difficulty with this explanation is that control personalities, although they claim to have lived on earth at one time, usually cannot be traced. (This is not always true: Lilian Bailey's control Wootton, was apparently an officer killed in the First World War.) Moreover, the very experienced trance medium, Mrs Blanche Cooper, is the one whose control passed on the information from Soal's fictitious character, John Ferguson, as though he were a genuine communicator from the next world.

Some trance psychics have told me that during trance they are quite unconscious. Others say that the 'guides' take them into the astral world. One psychic I knew felt himself to be a little behind his body, listening to every word spoken by the control, and he could discuss the communications afterwards with the sitter. (The others could not.)

I have carried out several experiments to see whether control personalities really can, as is suggested, 'see' both in the next world and in this world. For example, I was once part of an experiment in the SPR headquarters when we used mediums for a general extrasensory-perception experiment using Zener cards (a pack of twenty-five contains five of each kind — star, circle, square, cross and wavy lines). The cards were selected according to a random-number table in a room upstairs and presented to the agent, a bang being made on the floor to indicate to those in the library below that the card was being looked at. The psychic who was attempting to state what the cards were, was waiting in the library. She had a control who 'caused a small shining star' to appear at one of the five different cards (laid out on the table below) to indicate which was the card chosen in the room above. The psychic considered that her control went 'astrally' to the room above, looked at the card and then came back to the room below, indicating it by the star. The results were actually at the chance level — just as though they had been guessed in the ordinary way.

So what about the psychic's suggestion that we all have an astral body and that it is made of subtle material interpenetrating the physical and projecting all round? I carried out a number of experiments some years ago with psychics to test this particular hypothesis. But before I describe what I did let me say this. Two of the most remarkable psychics that I have known about were Annie Besant and Charles Leadbeater (see page 92). Leadbeater explained that Annie Besant was so spiritually developed that her aura extended many hundreds of yards all around. However, it was noticed that when Charles Leadbeater wished to know whether Dr Besant was in her office further along the corridor, it was necessary for him to walk along and see. Things are clearly not that simple!

But — the experiment! What I did was this. (I shall simplify a little without distorting the principles involved.) If the subtle body is interpenetrating the physical body and projecting all round then it should be possible to ascertain where the physical body is by observation of the projecting subtle body (the aura). So the experiment involved my covering up my physical body with a screen and asking the first psychic I tested whether she could still see my astral body. She said she could — by her clairvoyance. I then moved further behind the screen and asked the psychic whether she could still see my projecting astral body. 'No, I cannot,' she said. I then told her that I proposed to move near the edge of the screen (so that my astral body projected) and further behind the screen so that it would be obscured, and asked her whether she would be able to inform me of the position of my physical body from her observations of the astral. She stated that there would be no difficulty whatever in doing this; she did more difficult psychic tasks than that every day. So I moved to and fro behind the screen and listened to her statements: 'Now you are near the edge.... Now you are further behind.... Now you are near the edge....' and so on.

She had no difficulty in telling me, she considered, the position of my physical body relative to the edge of the screen even though she could not see it

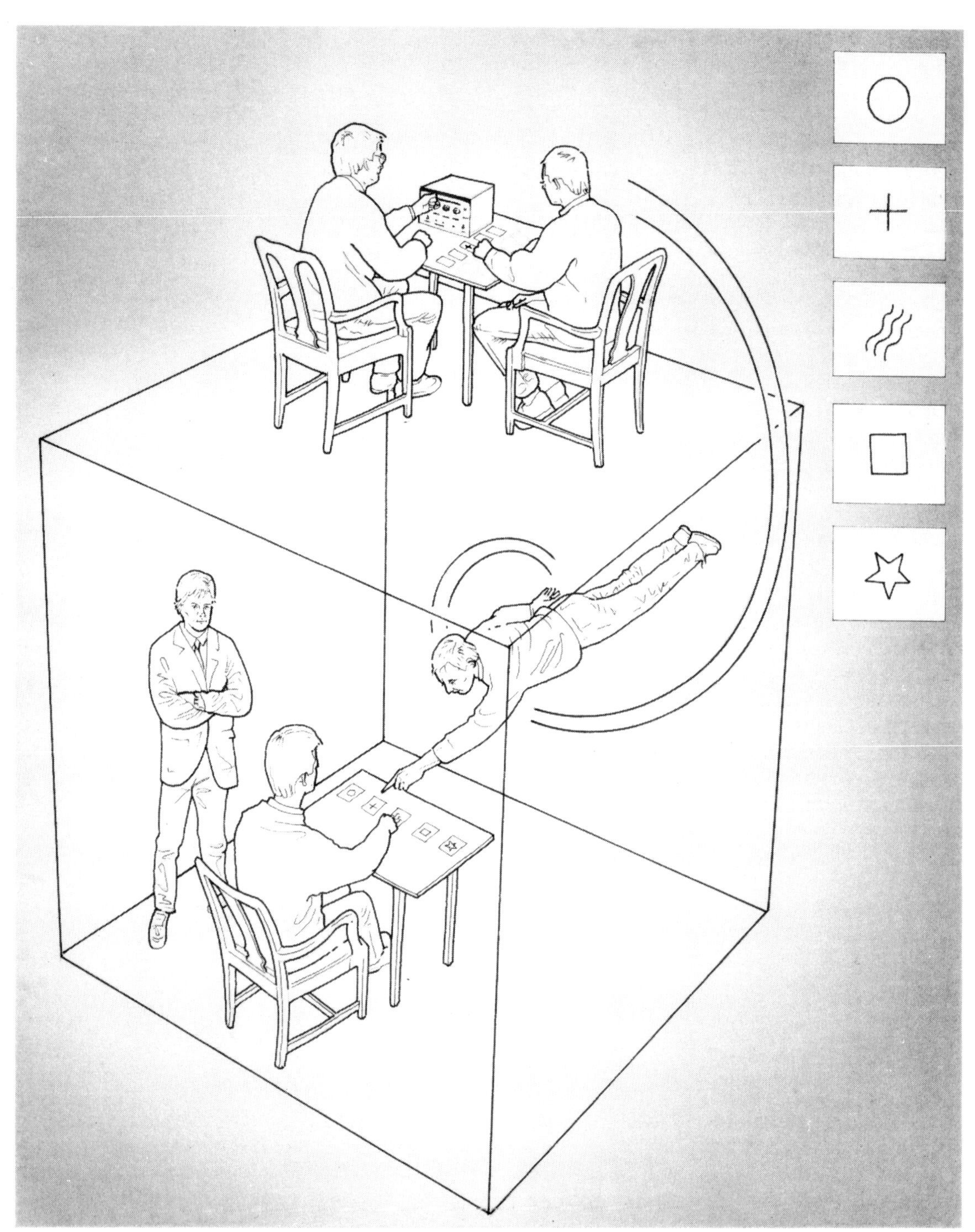

During an ESP experiment some people consider that a psychic's astral body somehow is able to view the cards. Is this what really happens?

in the ordinary way. Unfortunately her observations bore no relationship whatever to the position of my body relative to the edge of the screen. In fact I actually sat down on a chair after a while and listened to her calling out 'Now you are near the edge.... Now you are further behind....' I repeated this experiment with a fair number of other psychics, with exactly the same result.

So what do we understand from this experiment? The first point to be made is that it is clear that the position of the physical body in physical space cannot be determined by psychic observation. So why does the psychic consider that the astral body is interpenetrating the physical and projecting all round? The answer to that question is simply — because it appears to be so. She observes the physical body with her ordinary eyes in the ordinary way which the rest of us think we understand, and observes the astral body with her psychic sense also in the way to which she is accustomed. However, the astral body appears to be on the same centre line as the physical body only because it is associated with the same person and, not to put too fine a point on it, where else would it be?

The other level of consciousness of which the psychic is aware is clearly not interpenetrating physical space: this is a meaningless suggestion. So where is it? The answer is: it is not in physical space at all but in another space. Professor H.H. Price gave a good example to illustrate what is meant. He suggested that if you told him that last night you dreamed you were sitting under a palm-tree, and he asked you these questions: 'Where was that palm-tree? Was it at the foot of your bed or outside your bedroom door? And what was the height of the palm-tree? Was it five centimetres or thirty metres?' Clearly you would reply to him, 'These are meaningless questions. The palm-tree was not in physical space at all; it was in my dream space.' In just the same way the space observed in other levels of consciousness is a different space and not to be considered as having a particular relationship to the physical space. As will be seen in the last chapter, there is just as much mystery about physical space as there is about any of these other spaces. However, we are all so familiar with, and so naïve concerning, physical space that we often do not appreciate the difficulties until they are pointed out to us.

Does each of us have an astral body, made of subtle material which interpenetrates and projects from the physical body but which is visible to psychics only? (This may be referred to as the aura.)

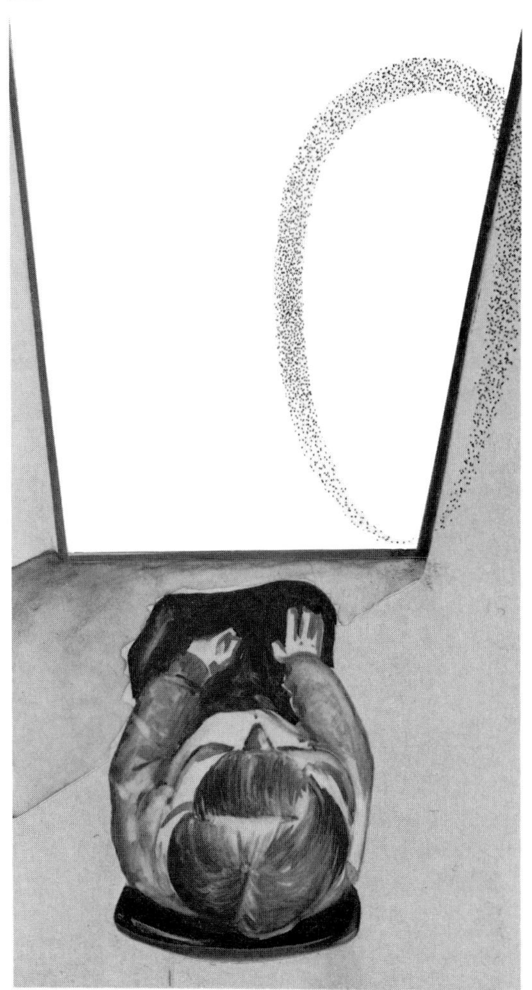

The aim of the experiment was to assess whether the psychic could ascertain my position behind the screen by her observation of my projecting astral body.

THE AURA

It was mentioned earlier that psychics who have 'trained themselves' appropriately can tell someone's mood by observing the colours of the aura. This, it seems to me as a result of many such discussions, is often true. It is difficult to be quite certain, of course, because there are all the normal clues available which are projected by one's physical appearance. Moreover, all psychics do not have exactly the same psychic observations of the subtler sides of other human beings.

A certain Dr W.J. Kilner many years ago considered that he could diagnose disease by observation of the 'aura'. He was referring to another 'body', the so-called etheric double, which is supposed in this theory to be different from the astral body, and to be made of a greyish 'semi-physical' substance again interpenetrating the physical body but projecting only some millimetres from its surface. This etheric double — smaller than the subtler astral body — was considered to be the 'mould' which assists the formation of the dense physical body and was also supposed to carry the vital energy of *prana*.

Dr Kilner suggested that he could sensitize his eyes and then observe the etheric double. The grey patches would indicate where the vital forces of life were not freely flowing and indicate to him where disease, either incipient or operative, was present. He did this by gazing at a light through a glass vessel containing a solution of dicyanin dye. After doing this he could gaze at the unclothed physical body of his patient and observe dark patches. He claimed to have success in diagnosing disease.

I carried out an experiment many years ago to test this and discovered that gazing at a light through that particular dye appears to tire the eye to the light over most of the middle of the spectrum so that when a body was then observed it was predominantly seen by the red and blue light at the opposite ends of the spectrum. Kilner evidently learned to focus a clear image of the patient's body surface. Presumably the light at one end of the spectrum gave him a blurred outline as a result of his focusing the light at the other end of the spectrum and the chromatic aberration of his eyes (the tendency of the eye to focus different colours at different points relative to the retina). So his blurred outline — the 'etheric body' — was in fact a blurred image of the ordinary physical body. So where did the dark patches come from to indicate the disease? If we assume that his diagnosis was often correct then we might also assume that Kilner's George was producing the dark patches in the right places from his knowledge of the disease present through ESP and his ordinary medical observations.

Generations of amateur psychical researchers have been persuaded to buy so-called Kilner goggles in order to see the 'aura'. My publication of papers giving the explanation seems to make little difference to the sales!

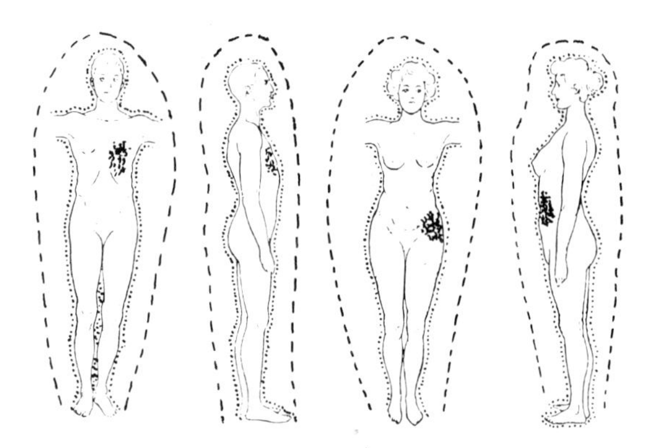

'Diseased' areas may be visible in the etheric double, surrounded by the astral body.

ASTRAL SPHERES AND REINCARNATION

Something which used to puzzle me greatly many years ago concerned what would naturally follow from the idea that the astral world, in which we find ourselves after death according to the psychics, is interpenetrating the physical, as are our astral bodies. It seemed to me that discarnate people living in such a world would see many 'unconscious' astral bodies belonging to people who were conscious in normal physical life. The teaching of Spiritualists' 'guides' concerning this appears to be that when we die our astral bodies normally float upwards to an astral plane which is some miles above the earth. It used to be customary to refer to different astral spheres in this connection. This is the same idea as the one involving hell, which is supposed to be a very low level astral region actually below the physical surface of the earth. In other words, going to hell was literally going down below. I put all this to a very distinguished psychic some years ago and suggested that, if that were true, we might actually be flying through our discarnate relatives when we took trips by jet plane in the stratosphere. He confirmed very firmly that this was indeed the case, in his view. Again, it is up to each one of us to make up his or her own mind but I must confess that this does not make a lot of sense to me. I prefer to think that these other spaces do not have any relationship to the physical world space.

A great many people believe in the general idea of reincarnation these days — and the scientific evidence that some people really have been reincarnated is, it seems to me, exceedingly good (see chapter 10). However, with regard to the experiences of psychics, they cannot know from their psychic perceptions whether reincarnation is true or not. Sometimes psychics have pseudo memories of other lives and they assume that these were other lives they themselves once lived. It is impossible to tell for certain whether this is true. As to the general beliefs in the Spiritualist community, a fair proportion of them believe in reincarnation and others appear not to. From the level of consciousness at which communication appears normally to take place it is clearly not possible to know. Some psychics speak of guides from very high levels who tell them about such matters but clearly a scientific person cannot merely accept this as evidence.

Spiritualists believe that after death our astral bodies may inhabit an astral plane a few miles above earth.

CONCLUSION

From this chapter you will have obtained some idea of what it is like to be psychic and the difficulties experienced by psychic children and their families. The 'gift' of psychic perception can be of great value if properly understood and developed, so it is sad that so few doctors understand normal healthy psychic perceptivity and that it is not studied by psychiatrists who could then perhaps better help psychotic patients. There is much valuable research to be done but meanwhile patients continue to be regarded as electrochemical machines despite the manifold inadequacies of this approach.

Most important, perhaps, is the value of the psychic's experience to philosophers: their concern is to understand the nature and limitations of our ideas of the physical world. Materialistic reductionist science is lacking in so many ways — and an appreciation of the psychic's viewpoint may well be an important missing element in our vision of the world.

8
DEVELOPING PSYCHIC POWERS

All the evidence suggests that the very best — the world-famous — psychics are those who are born psychic and who have anomalous experiences from their earliest years. However, some people have apparently become psychic as the result of an accident to the head. Yet others have had psychic experiences as a result of disease processes starting in the brain. Numerous psychics over the centuries, many of them religious contemplatives, have suffered from epilepsy. As might be expected, it seems perfectly clear that — as every mental state has a corresponding physiological state — so psychic functioning is associated with unusual brain functioning. Little is known as yet about this except that the evidence points so far to unusual phenomena in the right temporal cerebral region. Further reasearch may yet discover a great deal more about this.

IS EVERYBODY PSYCHIC?

But what of the rest of us? In my view everybody is psychic to a small degree. In the model of the mind represented by an iceberg (pages 11-12) the psychic would have a crack communicating readily between the unconscious and conscious regions of the mind. Perhaps the rest of us have minute crevices which might, with appropriate practice, be opened up into cracks. The evidence certainly seems to point that way. We shall have more to say about this on page 108. However, because of our rigid Western materialistic conditioning many people — especially those who like to consider themselves 'scientific' — tend firmly to dismiss any subtle psychic intimations; they will rely only on the information considered to be coming in 'reliably' (that is, via the five senses) combined with reason and logic. Perhaps many human possibilities of great value are lost as a result of this.

And what about the signs of psychism in children? The usual response when a child describes a psychic experience is along the lines, 'You naughty child! Don't make up such stories!' And what if a three or four-year-old announced that he or she had lived before in another town, in another country, and remembered it? Can you imagine what the average Western parent would say! Children perhaps occasionally come up with all sorts of intimations of immortality and we know only too well how to deal with them! In the case of a very psychic child (like Eileen Garrett) the child becomes secretive and withdrawn, mistrustful of adults. In most Western children, any tender psychic shoots are rapidly stamped into the ground — or deprived of nourishment until they wither and decay — and all is as though these psychic beginnings had never been. In any case most of the children who have reincarnation type memories have forgotten them bythe age of about seven.

Suppose we are eager to develop that extra window on the universe which is opened by the possession of psychism. There are many ways of developing a little psychic capability so that, at least in principle, we have some idea of the experiences of a good psychic. And there is no need to be shy about it. If friends do not understand we can always refrain from telling them! The way to encourage psychic functioning is to give it a chance and see what happens. The results can be quite surprising. I suppose my own friends would say that I was not in the least bit psychic but the reader will remember from earlier pages that I have had quite a number of interesting and illuminating psychic experiences as a result of following certain practices with a reasonable degree of persistence. Obviously, you are more likely to have effective results quickly if you are already some distance on the road; and the signs of psychic functioning referred to on pages 88 to 95 might be taken as an indication. However, if you can recollect no such signs there is no need to give up hope. Almost anyone can gain some little experience of the subject and it may prove very significant to them.

THE SLIDING GLASS AND OUIJA BOARD

Many people's first experience of psychic functioning is through the upturned glass or Ouija board: the experience can be frightening if it is not approached or interpreted correctly.

Many people have discovered a psychic amongst them as a result of playing the old parlour game with the upturned glass on a polished table and the letters of the alphabet in a circle around it, together with the words YES and NO. The fingers are placed gently on the bottom of the glass (and usually someone asks the traditional question, 'Is there anybody there?'). If you are lucky the glass will somewhat hesitantly move to the word YES. If it does not then a certain persistence will probably lead to success. Sentences can be spelled out and it is just as though there is somebody invisible who is present. Needless to say, this is not usually the case. George (or a combination of different Georges) can be very good at making up fictitious characters. There will be no problem in receiving answers to questions about the future. The only problem is that the answers are usually found to be false (see also page 89).

Sometimes, however, everything is of a very different order and quite good evidence of survival and/or telepathy is obtained in this way. I remember a friend of mine describing with some interest how he had stood by the mantelpiece watching such a session and found himself able to influence where the glass went by his thoughts. The same thing has happened to me during materialization séances, when characters clearly delineated in thought have appeared as materialized forms. How I wish there were available today some powerful physical mediums so that one could continue with good

The planchette has wheels or bearings so the pencil flows easily across the paper — to transmit words or messages in a successful experiment: it is sometimes used for automatic writing.

modern research, using all the instrumentation now available.

Somewhat similar to the upturned-glass experiment is the use of a Ouija board or planchette. A Ouija board is a polished board with the letters of the alphabet inscribed upon it together with the words YES and NO. It usually carries a pointed 'traveller' — a polished piece of wood, often with a soft felt under-surface, on which the fingertips are placed. The letters are indicated by the traveller in the same way as by the glass in the other system. The planchette is a small device of wood which is arranged on bearings or wheels and has means for carrying a pencil. The planchette is placed on a large piece of paper with the pencil point touching it, and the fingertips of the psychic are gently placed upon it. If success is achieved the planchette will write words upon the paper.

A natural development from the use of a planchette is automatic writing. In this practice the psychic holds a pencil on a piece of paper and the pencil writes words, the writer having no conscious control over what the pencil does. Sometimes the writers are in a state of slight dissociation, that is, a relaxed dreamy state: at other times they might be reading a book or listening to music. Probably fifty per cent of people could have success with automatic writing if they sat for perhaps one hour twice a week for a fair number of weeks. However, it would be most unlikely that the pencil would write anything of any interest. It might start off with nursery rhymes! In the case of some outstanding mediums (such as Geraldine Cummins or Hester Dowden) some most interesting 'communications' have been obtained by this means. There was, for instance, the occasion when Geraldine Cummins' pencil wrote the plot of a W. B. Yeats book in his presence (see page 45).

The dangers of experimenting in these ways are sometimes overstressed. Provided the approach is an interested, critical and scientific one, with a full understanding of the high likelihood of the production of fantasy rather than communications with the next world, there should be little danger. However, some years ago many children were developing psychiatric difficulties through practising with a Ouija board. Children have fertile imaginations and are easily frightened. Obviously such possibilities are present, especially if carried to excess and with the belief that any messages appearing are coming directly from people in the next world rather than from one's own unconscious. The approach and attitude of those concerned is all important.

PRECOGNITIVE DREAMS AND TABLE TILTING

An interesting experiment which almost anyone can carry out is a repeat of that described by Mr G.W. Dunne, relating to precognitive dreaming. Dunne describes in his book *Experiment With Time* (and also in *The Serial Universe*) how he wrote down his dreams immediately on awakening and discovered that sometimes they correctly foretold what was due to happen including, if I recall correctly, a dream involving looking at newspaper headlines of a paper which had not yet appeared. I tried this experiment myself for some months and my George very quickly grasped the idea of what I wanted and was waking me up every few minutes with a new dream to write down. I had very little sleep. So I was forced to abandon that particular experiment. If you do decide to try perhaps you should explain to George that you will experiment only every second or third night and put the pencil and paper beside the bed only on those nights.

Another common means of developing psychic powers or discovering whether they already exist is the well-known Spiritualist practice of 'table tilt-

Table tilting or table rapping may be used as a signalling method but a great deal of time and persistence is usually needed to produce results.

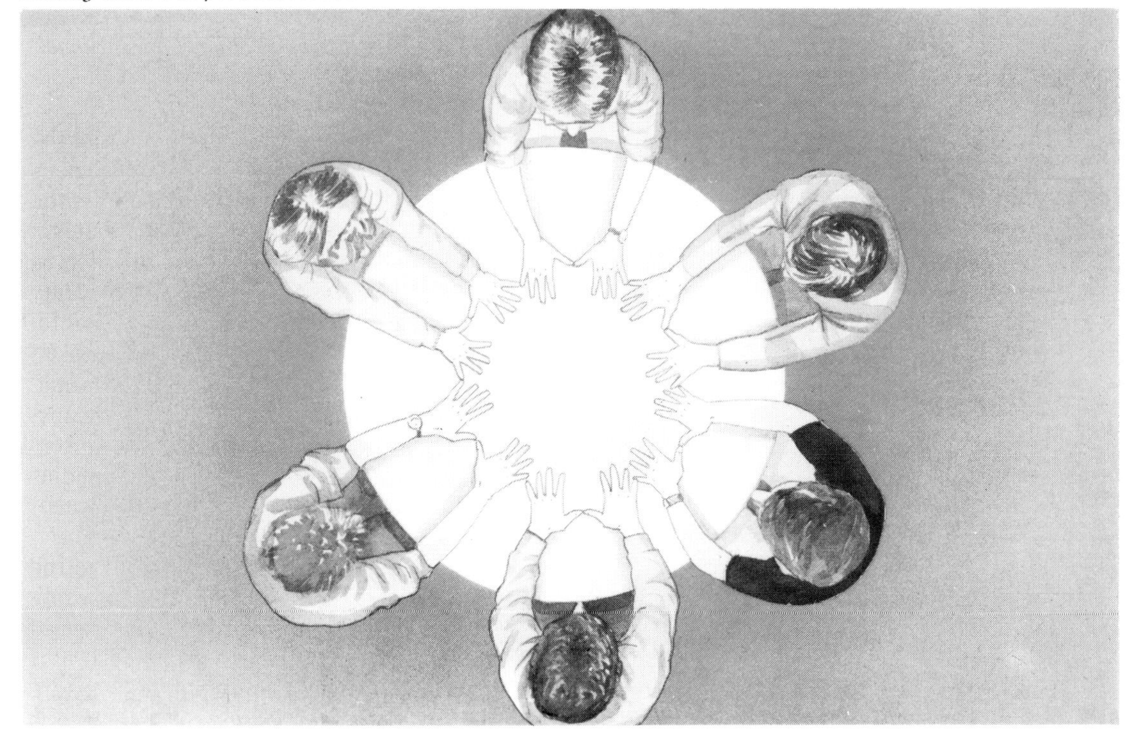

ing'. Several people sit around a table with their fingertips upon it and their thumbs touching and commence with the usual question, 'Is there anybody there?' Sometimes they merely sit without that stimulus. If they are lucky, after some weeks, or it may be months (a great deal of persistence is needed) the table will start paranormal tilting (with nobody pushing). Traditionally the messages are spelled out by the table tilting – once for letter A, two for B, and so on. It is all rather laborious! Alternatively, one tilt for YES, three for NO, and two for UNCERTAIN, is used. Even better, table rapping is sometimes the result of sitting and in this case a physical rap rather than a tilt is used as the signalling method. The old-fashioned, materialistic scientist would perhaps strongly deny that any such phenomena could possibly occur but I assure you that they are wrong. We cannot as yet understand such events but they most certainly do sometimes happen. (The so-called Philip Experiment, described on pages 143,144 and 145, very definitely proved this.)

SPIRITUALIST DEVELOPMENT

The Spiritualists help potential mediums to develop their powers by the use of the so-called developing circle in which the hopeful candidates sit in a circle, usually in a semi-darkened room, for an hour or so once a week. Some years ago I joined such a developing circle for the experience and so I am able to describe just what I observed. The group, consisting of about ten people, met every Saturday afternoon in a quiet room in the centre of London and sat upright on hard chairs in a circle. The medium who was directing the circle started off each session with a short talk which, in effect, amounted to suggestion. She explained that the participants should sit with their spines erect, bodies relaxed and eyes closed, and 'realize' that they were being worked upon by 'unseen operators' in the next world. It was explained that there would be various feelings, sometimes of a cap being put over the head or of pressure on the forehead. All this would be evidence of the actions of the unseen operators. Any images or impressions which came to the sitters during the period of development (lasting about one hour on each occasion) should be reported at the end of the session.

It is obvious that sitting in a semi-darkened room, quiet and relaxed, and imagining that one was being worked upon by unseen operators would lead most people to unusual sensations. I certainly had feelings of pressure on the forehead and felt as though a 'psychic developing cap' had been placed upon my head. However, I had few if any other experiences. Some of the other participants described various characters they stated they could see clairvoyantly and reported a selection of Chinese or Red Indian guides working on their mediums.

Sitting in a relaxed state for an hour or so every week imagining that one is being operated upon by perhaps a Chinese or Indian guide is very likely to lead ultimately to the creation of such a secondary personality. Thought and belief are exceedingly creative and it would not be at all surprising if a sensitive person and an incipient psychic developed into a trance medium with a Chinese or Indian control. In one sense this is of course self-deception but in another sense the creation of this psychic machinery might well act as a vehicle for the transfer of ESP information acquired by George. It would be unreasonable to state categorically that the development of a secondary personality (and we must certainly not rule out the possibility that the control personality may sometimes be a distinctly different personality from the psychic) is a useless system of self-delusion. This by no means follows. However, it is important to try and obtain an unbiased and scientific assessment of what actually takes place, together with some estimate of its value.

RELAXATION/MEDITATION

Meditation has been mentioned earlier as, according to Indian tradition, leading to the psychic powers which develop on the way to complete unity or yoga with the divine source of all life. Many people who have practised meditation in the West-and relaxation is a precursor to meditation - have had psychic experiences. I developed some of my own psychic experiences as a result of what might be called straight relaxation and holding the mind almost blank. In effect this is the system of transcendental meditation (or TM), the mind being assisted to become blank by the repetition of a meaningless word or phrase, that is, a mantra. (All mantras are not meaningless, but those used for TM usually are.) The aim in TM is to put oneself in tune with the infinite source of all possibilities, that is, to transcend the rational mind.

I practised TM (having been initially instructed in the normal way by an expert) for a complete year and I had no transcendental experiences of any kind. As a result of my earlier long practice of different types of other meditation and relaxation I had no problem in holding my mind clear of almost all thought but it sadly led to nothing and I abandoned the experiment after one year. In fact, my acquirement of this experience of TM was primarily because I had read a number of papers indicating ostensibly valuable physiological and biochemical side-effects of the practice. Holding the mind free of thought and the relaxed body free of stress appears to lead to the complete normalization of all bodily systems. This is claimed to lead to a cessation of many unhealthy and undesirable characteristics like the craving for drugs of various kinds such as nicotine, alcohol and others even more harmful.

My aim was to carry out a research programme with Governmental support to see whether much stress — and consequential illness and expense — might be avoided by the regular practice of meditation. The aim of the work was to check the TM claims by using control groups of people as well. These would relax, with or without listening to music they liked, to see whether the physiological effects were unique to TM or whether they would follow from merely sitting and relaxing twice a day for two periods of twenty minutes. The potential savings to the National Health Service in Britain and corresponding savings in other countries would be enormous. Unfortunately my collaborator (a neuropsychiatrist) and I were quite unable to obtain any Governmental support — owing to the demands for all the research money that was available (in the UK) to support areas of research better understood by the grant-giving bodies! So that research is still to be done.

BIO-FEEDBACK/SELF-REGULATION

It has been shown that it is quite possible to bring under conscious control any automatically functioning part of the body, provided that the function to be controlled can be made overt. For example, the right hand can be made warmer and the left hand at the same time made cooler if the temperatures of the hands are measured and the difference in temperature is indicated to the subject. A meter or perhaps a tone changing in pitch, can be used to indicate the difference in temperature. Similarly the forehead can be made cooler, the heart rate slowed down or speeded up, or the blood pressure changed.

In order to produce a differential temperature between the hands the subject will use the creative power of the imagination, the details of the most effective method probably depending on the person concerned. One subject might find it occurs when relaxing the body and mind and imagining that one hand is holding a block of ice while the other is put into a bowl of hot water. Raynaud's disease in which, for example, one finger becomes cold and white, can be cured in this way by attaching a small thermometer or thermocouple to the finger and combining relaxation with imagination. In a similar way epileptic attacks can be greatly reduced or eliminated and tension headaches may be removed by learning to control the temperature of the forehead or the tension in the muscles of that region.

BRAIN RHYTHMS

Autogenetic training, assisted by bio-feedback, is particularly important to mention here because it can be used as a means of learning to achieve states of consciousness which are likely to lead to psychic experiences. It is well known that the electrical activity of the brain as measured by the electroencephalograph (EEG machine), can be observed by sticking electrodes to the surface of the scalp. These brain rhythms change with states of consciousness; they are fairly rapid during normal, waking thinking periods and slow down with relaxation, reaching exceedingly low frequencies during deep sleep. The state of reverie, half-way between waking and sleeping, which leads to hypnagogic imagery (described on page 85), is associated with fairly slow brain rhythms referred to as theta rhythms.

Actually, the brain waves have been roughly divided into four bands known as beta, alpha, theta and delta rhythms. Delta rhythms range from about 0.5 Hz (cycles per second) to 4 Hz (subjects producing delta in any significant amount are generally found to be asleep or otherwise unconscious). The next higher band, from 4-8 Hz is theta — these theta rhythms usually being associated with the near-unconscious or subliminally conscious states, often appearing with drowsiness. Alpha rhythms, from 8-13 Hz, are associated with a state of greater awareness than those of theta and often occur when people close their eyes for a short period. When a subject is active with eyes open, the EEG record generally shows the presence of beta waves ranging from 13-26 Hz or higher. Usually more than one range of frequencies is to be found and, when the eyes are closed and the body relaxed, alpha waves usually begin to appear, being mixed with beta. As time passes the alpha becomes smoother and more regular in frequency although the amplitude fluctuates. Theta waves begin to appear as the body becomes much more deeply relaxed and the subject becomes drowsy. In meditation a rather smooth theta pattern is often correlated with a relaxed body, quiet emotions and a tranquil mind but consciousness does not diminish.

It is quite easy to indicate — by electrodes connected to some simple electrical equipment — the presence of brain rhythms in these particular bands in different parts of the scalp. For example, if the frequencies are multiplied by about two hundred (easily done electrically) then they become audible as musical notes, and the presence or otherwise of particular frequencies can be recognized while relaxing and/or meditating with closed eyes.

By using bio-feedback in this way it is possible greatly to speed up the achievement of appropriate states of consciousness in meditation which are likely to lead to psychic experiences. The traditional Eastern view is that raja yoga practice (meditation) leads to psychic faculties.

THE GREEN EXPERIMENTS

Dr Elmer and Alyce Green (friends of mine at The Menniger Foundation in Kansas) pioneered the use of bio-feedback in connection with the so-called autogenic training of Schultz. (In the early 1900s Schultz developed in Germany the use of auto-suggestion and visual imagination to produce deep relaxation and the normalization of a number of bodily disorders.) The Greens have measured the brain rhythms of a number of yogis, some of whom were able to perform psychic tasks. Their most accomplished subject was a certain Swami Rama from whom they learned a great deal concerning brain rhythms and other physiological parameters associated with his meditative practices. Swami Rama was able, without harm, to cause his heart to go into a state of fibrillation — an irregular rapid fluttering — and then return it to normal. He could also produce considerable differences in temperature between the left and right sides of one of his palms. In addition, Swami Rama was able, under strict laboratory-test conditions, to cause movements of a small pointer by psychokinesis.

There are clear correlations between mental states and physiological changes. It would obviously be greatly helpful therefore, when attempting to follow a yoga teacher's instructions on your mental activity, to be able to observe the physiological changes. This might, and indeed does, greatly speed

This is how the room was arranged during the psychokinesis experiments with Swami Rama.

up the process, as compared with the traditional way of merely trying without any sort of information feedback. Swami Rama himself, when he returned to India, asked the Greens to let him have some biofeedback devices to indicate brain rhythms for, as he put it, 'training young yogis and eliminating fakers'. (There are many confidence tricksters who are not real yogis to be found in India and some of them have enormous followings.)

It is obviously not possible to follow an advanced yogi into the intricacies of his mental states by using physiological correlates alone. However, it is clearly helpful to speed up the start of the journey!

Some rather different psychic capabilities were observed by the Greens when they studied a well-known Western Sufi, Jack Schwarz. (Sufism is the esoteric side of Islam.) Jack Schwarz's most remarkable demonstration was of insensitivity to pain and the control of bleeding. He demonstrated pushing steel sail-maker's needles through his biceps and preventing bleeding. Evidently he also prevented infection because there was no normal sterilization of the needle.

Dr Green explained that Schwarz produced the normal beta waves all the time he was sitting in the laboratory but when he put the tip of the needle to his biceps he began to show alpha waves. This is the reverse of what would happen to a normal person. The explanation of this appears to be what many of us have noticed in a small way, that when pain is experienced, if the attention is turned away from it on to something else, then it is noticed very much less. Some dentists have distracting stimuli in their treatment room for exactly this reason.

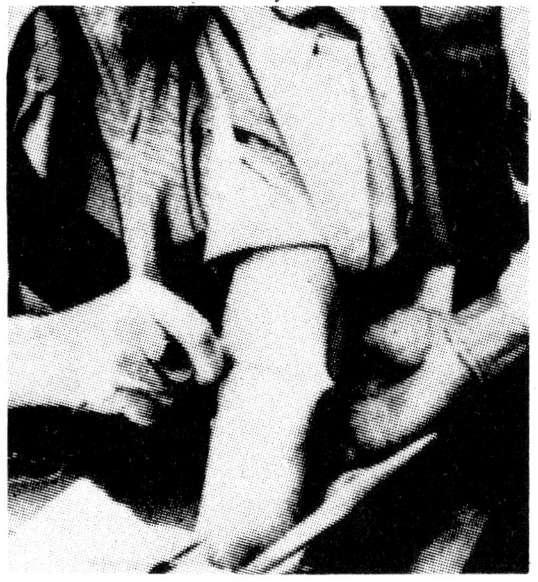

Jack Schwarz drove the sail maker's needle through his arm apparently quite painlessly and without any bleeding.

When Schwarz began pulling the needle out of the other side of his bicep there was no bleeding and the puncture holes closed up immediately, leaving almost no trace. However, when Dr Green interrupted Schwarz as he was pulling out the needle by asking whether the wound was going to bleed, it actually did bleed, his concentration having been destroyed, but he was able to stop the bleeding at will almost immediately. In this demonstration Jack Schwarz explained that he thought of his arm as an object not attached to him and that detachment is clearly linked with the production of alpha waves.

With regard to the control of infection, Jack Schwarz gave an answer similar to that of Swami Rama. He explained that if the body (meaning George) understands that it is not to interact with, or react to, any foreign material, infection cannot commence. Pain and bleeding can be controlled, of course, by using hypnosis, provided the subject is able to go into a deep enough trance. A number of doctors use this method. Clearly (using our model of the mind) the communication with George is by hypnosis from the doctor. In the case of yogis, such as the two who have just been described, they have themselves trained George to do what is necessary and are consciously able to control the direction of their thoughts. In other words, the conscious mind is put into that blank or relaxed state when it will not notice the pain.

Another demonstration which Jack Schwarz could do consciously at will was to anaesthetize his body so that he was able to hold a burning cigarette against his skin without any physiological indications of pain. No blisters formed but the top layer of skin was burned. Again, this can be done, and frequently is, using hypnosis. Jack Schwarz had to 'ask his body for permission' before carrying out experiments. That surely means communicating with his George, who is running the bodily systems and will know if a test can be carried out without harm.

The control of pain when wounded is of course only one aspect of the problem with which one is faced. The other aspect is to produce healing as quickly as possible. This can be done by appropriate use of clear visualization techniques, that is, by telling George exactly what to do. (Jack Schwarz uses this method also.) This will be further discussed in chapter 9, *Paranormal Healing*.

A final word about Jack Schwarz: he claimed that he was able to see auras whenever he wished and was thereby able to name specific diseases within the subjects he observed. He learned this by experience as a result of seeing the auras of many people who were suffering from various diseases.

CONCLUSIONS

This is not a book about yoga. There are many hundreds of different kinds of yoga in the two major areas of *hatha yoga*, involving bodily exercises and posture (asanas) and *raja yoga*, involving the use of the mind and the way that life is generally lived. Anyone who wishes to learn about and practise yoga would be well advised to study the many excellent books on the subject and find which type of yoga appeals most. For worthwhile results it is hardly necessary to say that some persistence is needed.

I hope very much that your reading of this chapter will produce several valuable results.

For instance, if you know a family who has a psychic child you might be able to recognize it as such and not consider that the child is making up stories or, worse, is mentally ill. Psychic children should be helped to grow up happy and healthy — despite the general lack of understanding in our very materialistic society.

If you decide to try the Ouija board or upturned-glass experiments — you will know not to take it too seriously. Impressionable people have sometimes been frightened by the results but being aware of what George is doing should allow you to develop a calm attitude and a sense of proportion. Few will have the tenacity to achieve results from table tilting or rapping — it might take a year or two. If you try — good luck! But make it all enjoyable.

Perhaps the major conclusion to be drawn with scientific objectivity is this: if you make efforts via concentration and meditation to collaborate with George and explore the unconscious mind (with or without the help of bio-feedback), you will learn how to make the conscious mind tranquil and still. The first fruit of this — if it is done regularly for say twenty minutes, twice a day — will be a tendency for all your bodily electro-chemical systems to return to healthy normality. (Chapter 9 contains further data.) Although meditation is not a cure-all, a tranquil mind tends to lead to a healthy body. Stress (and its bodily chemical products) serves a purpose in moments of danger — but not when you are going to sleep at night!

9

PARANORMAL HEALING

TODAY'S LIFESTYLE

From what has been said already it will be very clear that a human being is a great deal more than an electrochemical machine. All human beings are linked together, parts of a whole, and that whole transcends time and space. The enormous importance of belief and the creative power of thought have been pointed out. But what does an ordinary general practitioner do when presented with a patient? Almost all the GP's training has been of the reductionist kind in which the patient is assumed to be an electrochemical machine: though psychological factors are sometimes mentioned they certainly do not occur very often in any first degree medical course. Certainly, within the over-burdened National Health Service of the United Kingdom, a doctor has little more than five to seven minutes with each patient. There is scarcely opportunity to wish the patient the time of day before the doctor is reaching for a prescription pad to write out a prescription for some chemicals which will suppress any symptoms the patient is presenting. If the patient is clearly in a worried disturbed state then tranquillizers will probably be prescribed.

Clearly this description is only a very rough approximation to reality and cannot apply to every case but it will certainly be recognized by many patients. Tranquillizers and other similar drugs are a multimillion business these days.

But why are most of these patients ill in the first place? Human beings are a psychosomatic unity of mind, emotions and body. Most people do not know how to cope with life. (In the United Kingdom, doctors commit suicide about twice as often as do other groups!) Only a small percentage of the population

When a patient is in a disturbed or anxious state tranquillizers will often be described. These may suppress symptoms but will not deal with the root causes of anxiety.

take religion seriously: it has been almost destroyed by Western materialistic science. People feel helpless in an alien universe and very far from safe and secure. Scientific and engineering advances have given the majority of people in the West comfortable lives and insecure helpless minds, filled with stress. It is highly probable that a good many diseases are the result of this: the struggle to 'get on' and keep up. If the stomach pains, arthritis, ulcers and headaches are treated chemically, only the physical side, the symptoms, are being treated. However, the prime cause is probably a mental one. The patients need wise philosophical advice on how to live their lives: a little philosophy, a little reassurance that somebody cares. But who has time — or is wise enough — to do this? The former source of guidance, the Church, has been practically destroyed by science and science has put nothing in its place.

What I am suggesting (and most people, including many doctors, probably know this as well as I do) is that a majority of diseases are probably psychogenetic, that is, produced by the mind. The milder ones can be cured by a bottle of coloured water and a strong suggestion that the patient will get better. The more serious ones need a complete change in lifestyle, to reduce stress and give tranquillity and equilibrium, combined, of course, with a balanced diet, fresh air and exercise. Probably some seventy per cent or so of the diseases with which a general practitioner has to deal are psychogenetic in origin and will probably improve anyway, whatever is done — even if it is nothing.

Unorthodox healers tend to have more time to talk quietly to their patients and give advice, and this advice is likely to be accepted, particularly if paranormal sources are claimed for it. All this can lead to a more tranquil mind and body, and natural healing, whether or not the healer's claims are true.

As a result of all this, many more patients are visiting unorthodox healers these days. The figures for the United Kingdom indicate that eight to ten per cent of all medical consultations are with complementary practitioners.

It has already been pointed out that George in each person's unconscious is not rational and logical but a-logical. George will sometimes acquire all sorts of inappropriate — indeed downright foolish — ideas and create general havoc with the physical body which he normally organizes so smoothly and efficiently. A case was mentioned on page 12 where George linked going on to a stage with a freezing up of the vocal chords so that a talented soprano was unable to sing a note in front of an audience. If such occurrences take place the subject's psychiatrist will be, amongst other things, very interested in the production of dreams because these often provide a clue to what George is doing. The commonest mistake which George makes is to keep the subject in a continuous state of 'fight or flight' where adrenalin is being pumped into the bloodstream ready for action; but in the artificial, self-controlled state that we are expected to maintain in our Western civilization such physical release is rarely required or permissible.

TWO CASES CURED

Sometimes a health problem caused by George can be put right by direct instruction to him. I remember well two cases in my own experience when doctors, knowing of my interest in hypnotism, asked me to assist. (The ordinary GP in the consulting room, having such a short time with each patient, rarely has time to use hypnosis — even if he or she knows anything about it, which is not too often.) In the instances concerned, I effected cures by hypnosis which relieved one case of migraine and one of insomnia.

In the case of the migraine it was necessary to see what could be done in one weekend as the patient lived in another part of the country. This particular patient suffered one chronic disabling migraine attack about every three weeks, and had been incapacitated by these for many years. She proved to be a reasonably good hypnosis subject and rapidly went into a medium-to-deep trance. I suggested to her George that she would in future remain calm, tranquil and relaxed during her normal daily life and the migraine attacks would, consequently, gradually fade away. I gave her these strong suggestions during three separate half-hour sessions on a Saturday and a Sunday and then had to leave. Thereafter she had no migraine attacks at all (this is a good many years ago) and had only one slight headache some months later. She was able to throw away all her medication.

In the case of the insomniac, she lived within easy travelling distance of where I live and I was able to give her treatment for about half an hour, once a week for about five weeks. The treatment in effect consisted of exactly the same: that she would remain tranquil and relaxed during most of her waking hours and gradually commence sleeping soundly at night. (Most people remain full of stress during most of their waking hours and during a good deal of their sleeping hours as well.) The result of this treatment was that she too was able after several weeks to sleep well with no medication and, having visited over several years three hospitals to no avail, was able to throw away all her drugs.

It seems sad to me, after my experience of both these cases (and the reader will appreciate that we must not draw general conclusions from so few cases) that hypnosis plays very little if any part in the vast majority of undergraduate medical courses.

UNORTHODOX HEALERS

It will not be appropriate within the context of this book to devote space to the methods of holistic healing referred to generally as alternative medicine or, better, complementary medicine. Several excellent books describing these methods have been published lately. (By complementary medicine we refer to homeopathy, osteopathy, chiropractic, acupuncture and herbal medicine.) Some of these disciplines have a professional training lasting several years and a proper system of examination and qualification. In addition they all tend to be what I have referred to as holistic, that is, the practitioner strongly realizes, as a basis of his or her profession, that *a human being is a combination of mind, emotions and physical body and all these react upon each other.* The practitioner therefore spends time looking at the other areas of the patient's life and not only at the symptoms presented.

The enormously creative power of the mind and belief is considered of great importance in some of these disciplines, although not in all of them, and not by all practitioners. Some of them seem to have very unscientific views as to how they work, that is, their model representing the system appears to a scientific person to be rather naïve. In view of the holistic nature of the practices it is clearly not possible, in doing research on them, to use such orthodox methods as double-blind cross-over tests such as are used in the testing of drugs in the orthodox allopathic medical systems. It is appreciated by unorthodox practitioners — or at least by many of them — that the mind of the practitioner as well as that of the patient is an important component in the system and certainly cannot be eliminated.

By contrast, reductionist allopathic medicine views the patient primarily as an electrochemical machine, and the doctor is therefore looking at the bit (or bits) that have broken down with the aim of repairing the breakdown chemically or surgically. The healing propensities of the patient's own mind hardly enter into this; in fact many patients behave with their doctors as if they have no responsibility for their own health and it is up to the doctor to cure them. This is directly contrary to most of the complementary practitioners' views, in which *the patients have a vital responsibility for the restoration and maintenance of their own health.*

Let us therefore now consider what this chapter is primarily to be about, that is the sort of healing methods which go on in 'healing sanctuaries', or their equivalent. There seems to be a much closer relationship to what we have been calling the paranormal in these healing methods.

Meditation

Transcendental meditation (TM) has already been mentioned. It is not primarily a method of healing but the result of regular meditation (of almost any kind) appears frequently to be the clearing up of all sorts of health problems as the whole of the bodily systems become more tranquil and balanced as a result of the tranquillity of mind produced. Stress is greatly reduced as indicated by all the usual means (measured in connection with TM): cholesterol, adrenalin, heart rate, blood pressures and electrical skin resistance (the latter indicating tension of the autonomic nervous system). All the methods of raja yoga (meditation) help in these directions. Usually there is a philosophical side which puts the meditator into a satisfactory relationship with the rest of life and free of unwarranted stress. Health is often improved by correcting stressful mental conditions — especially, of course, at the unconscious level.

Hypnosis

Hypnotic suggestion, mentioned earlier, is the commonest (almost obvious) way of changing conditions in the mind at the unconscious level, that is, the beliefs and practices of George. This is still perhaps an unorthodox method of healing as it is certainly not in the armoury of most medical practitioners — who have insufficient time to use it anyhow. I have already mentioned (on page 113) the use of hypnosis in parapsychological research and its use to cure a chronic migraine and a chronic insomnia case. Suggestion under hypnosis is important to mention further because suggestion may be the most important component — if not the only component — of a number of healing methods. (The 'healer' may have quite different ideas of course.) A tranquil state of mind and body plus a strong belief with appropriate suggestion is often exceedingly effective, whether or not the belief is correct. (As we are considering the importance and creative power of the mind on the body the word correct is a somewhat slippery one.)

Let us have a few examples. In Christian Science there have been many claims of sudden healing as a result of reading Mary Baker Eddy's *Science and Health with Key to the Scriptures.* Whether or not mind is the only reality — which is the teaching in this instance — if the subject strongly believes that it is, the healing can sometimes take place.

The sort of faith healing which occurs in the Catholic community at Lourdes is also of course closely related to belief.

Hypnosis is a powerful tool: it allows a psychiatrist or a healer to make direct contact with 'George' and to influence the belief structure of the subject. An interesting example of this is discussed on page 43.

PSYCHIC HEALING

I have always been interested in the kind of healing which goes on in the Spiritualist community. I particularly examined the methods of the well-known British psychic healer Harry Edwards, some years ago. Certainly I observed there a number of occurrences which were not easily explained, or should I rather say, which orthodox doctors could not easily explain within their own framework. I once observed a lady come on to the stage in the Royal Festival Hall in London where she revealed a large projecting goitre in her neck. One of the healers working with Harry Edwards gently stroked it for some minutes and it apparently disappeared completely. I could see a quite different silhouette against the light and a doctor with whom I discussed the matter told me that there was certainly nowhere to which the healer could have pushed the goitre out of sight.

Of even greater interest was Harry Edwards' treatment of a number of arthritic old ladies who came up on to the stage scarcely able to hobble and with their finger joints very swollen and stiff. I watched Harry Edwards place his hands against the backs of the patients and straighten them up; then he continued by manipulating their fingers. He took each finger in turn and moved it about until it seemed to be free. At the end of the process the patients appeared to be able to move their fingers in an almost natural way and left the stage walking practically normally. I should mention that Harry Edwards considered that the deceased doctor Lord Lister worked 'through him'. That is, he considered himself to have collaborators in the spirit world.

I discussed this remarkable phenomenon of the apparent removal of the symptoms of arthritis with a distinguished psychiatrist. He told me that he had observed exactly the same procedure and results. Moreover, at the end of one healing session, when the healers had gone, an arthritic old lady had hobbled hurriedly into the hall saying to him 'Oh dear! Am I too late? Is it all over? My train was terribly late.' The psychiatrist answered, 'I am afraid that is so, but I am a healer myself. Would you like me to help you?' The patient said that she certainly would so the doctor sat her down on a chair in the hall and he did exactly what he had seen Harry Edwards do. He straightened out her back by gentle pressure and he freed each of her arthritic fingers one after the other in the same way. That doctor had no belief whatsoever in 'spirit collaborators' but he was able to do exactly the same as the healer. He told me that he most certainly could not have had that effect in his own hospital when wearing a white coat and being called doctor.

I am sometimes asked how long these paranormal healings last. I must say that I have no idea. Not having had a medical training, I do not feel qualified to pursue such cases in a way which would be of any medical value.

Following the experiences I have just described led me in due course to a large Spiritualist organization which had regular healing sessions. I telephoned the chief healer and, as is usual, she was kind enough readily to give me permission to sit in and observe. I went along one evening at the agreed time and when I appeared was asked to sit in a chair at the end of the row of patients. I watched the healer, who was working in trance; she considered that the treatment was being undertaken by a discarnate control. Her results appeared to be quite interesting but I was medically unqualified to assess them. However, when all the patients had been dealt with I was myself asked by the healer in trance to come forward and sit on her stool. She asked me to remove my jacket and then placed one hand at the top and the other at the bottom of my spine. I had no idea what to expect but the result of this was most peculiar. The feeling I had was exactly like that of warm water running down my back. After several minutes of this treatment the healer asked me whether I felt any better. 'Actually', I said, 'I am perfectly fit and only here to observe.' The healer then said to me, 'My control put this on as a demonstration for you.'

Some time later I studied another psychic healer who, when I was there, was treating an arthritic old lady. The healer put her hands, one on each side, on the patient's knee and the patient told me that she could feel the knee getting hotter and hotter and she felt this to be of considerable benefit. I placed beneath the healer's hands a thermocouple and observed the temperature at the place where the old

Photograph courtesy of The Harry Edwards Spiritual Healing Sanctuary

Many people have witnessed Harry Edwards performing psychic healing at public meetings.

lady felt the temperature to be rising. The meter indicated no change in temperature. I then asked the healer if she would allow me to carry out an experiment. She was very willing, so I asked her to sit to one side and allow me to perform. I wrapped several turns of coloured wire round the patient's arthritic knee and attached the ends to a decade resistance box. (This is a rectangular box carrying a number of graduated knobs and containing only electrical resistance between its two terminals. The knobs are adjusted to vary the resistance. The box contains no battery or other source of supply and is quite inert.) I attached the wires to the two terminals and twiddled the knobs, looking very wise and scientific. I asked the old lady whether she could feel anything. 'Yes, I certainly can,' she said, 'I can feel my knee getting hotter and hotter.' She felt exactly the same benefit as she had obtained from the healer and yet, so far as I was concerned, I was doing absolutely nothing at all except provide a strong suggestion.

Something which must be mentioned in connection with psychic healing is this: patients often have told me that they were cured of cancer by some psychic method. When I enquire rather more carefully into the details I discover that they perhaps had abdominal pain and the healer both diagnosed the cancer and cured it by psychic treatment. The sceptic will appreciate that there was no proof of any cancer being there in the first place and the pain may merely have been due to indigestion, which would have disappeared in due course anyhow. Again, I must emphasize that psychic healers are, most of them, transparently honest and genuine people with goodwill towards their patients. However, their explanations and the terms by which they describe their healings sometimes do not stand very much scientific analysis. Nonetheless, why should that matter? If the patient is healed, splendid! However, if they go around for the rest of their lives telling all their friends that a particular healer cured them of cancer then a number of people may be grossly misled and could refrain from visiting their normal doctors when it may sometimes be rather important for them to do so.

AN EXPERIMENT WITH SUGGESTION

Some psychic healers do not claim discarnate help but think of themselves as channels for forces unknown to science. I have noticed that they often put their hands in the right places on the patient almost at once. (This may of course be the result of slight indications from the patient which they recognize.) I remember well my visit to a well-known healer of this kind, an event which turned out to be particularly interesting. I was most warmly welcomed: I have already mentioned several times how eager most healers are to have scientists examine their methods, which they consider are grossly neglected by the medical profession. I observed the healer's treatment of the particular patient who was there at the time and whose ankle had been badly injured by a terrorist bomb planted in London. He had undergone a number of operations to his ankle which was now heavily scarred, and he was able to move his foot to the left and right through only a very small angle. When the healer put her hands on either side of the ankle (without touching it) the patient experienced strong and growing movements of his foot which moved to left and right through a very much greater angle than before — to his great discomfort. He winced with the pain and could feel discomfort to the top of his leg. The healer told me that she looked upon herself as a channel for forces unknown to science which came to her from she knew not where and flowed through the patient via her hands. 'It is', she said, 'up to you scientists to find out.' I then suggested that we carry out an experiment. This we did as follows.

I said to the patient, 'Let me see if I can do that.' I put my own hands on each side of the injured ankle, exactly as I had seen the healer do, and I produced absolutely no result. The patient felt nothing. 'Sorry!' he said, 'You are just not a healer.'

I then explained the experiment I proposed to carry out and everyone agreed to participate: I said that I would write down on a piece of paper the words HEALER, MYSELF and NO ONE in random order. I covered a large piece of paper with these three possibilities in what seemed to me to be an entirely random order. I explained then to the patient that I proposed to blindfold him (which I did with a scarf so that he had no possibility whatever

With the patient blindfolded, the healer and I (and no-one) put our hands near his ankle in a random order.

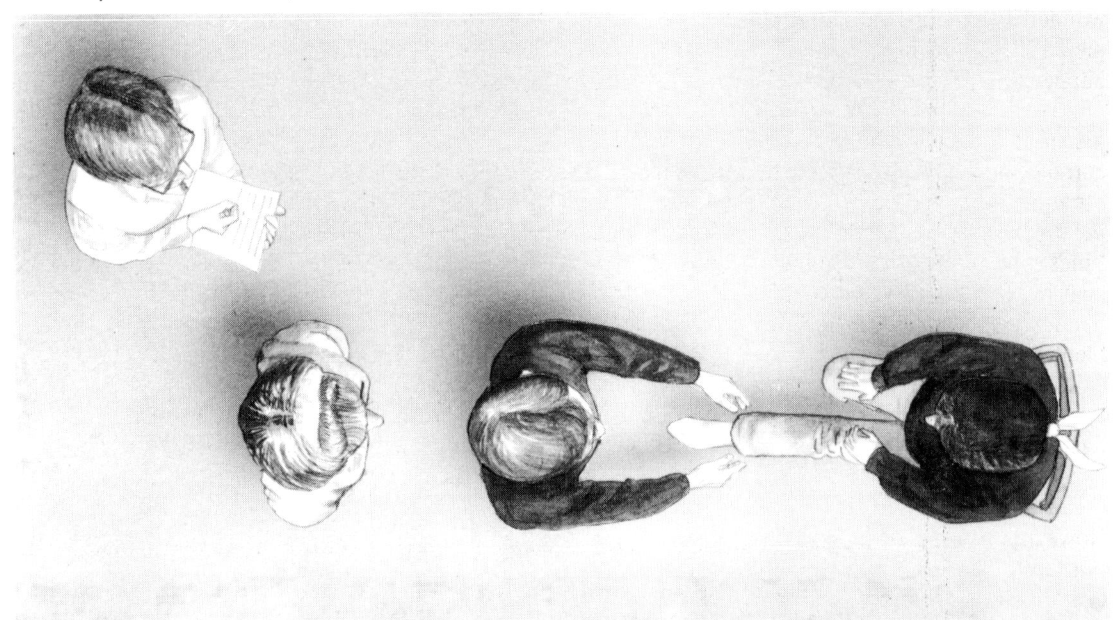

of seeing) and that I would ask a young nurse, who had joined us in order to help, to point to the next person whose name appeared on the list. That person would walk up to the patient and put their hands on either side of the injured ankle. If the next person was 'no one', then I would walk up but put my hands behind me. It was organized so that the patient would hear exactly the same sounds on each occasion. He was asked to tell me whether it was the healer, myself or no one whose hands were near his ankle. He considered that he would have no difficulty whatever in distinguishing between the healer, and myself or no one, but — as my own results were to be equivalent to 'no one' — he did not think he would be able to distinguish between these last two categories.

So we did the experiment. The nurse pointed to me and I walked up and put my hands on either side of the ankle. She then pointed to the healer, who did the same. This was followed by the nurse's giving a shrug of her shoulders to indicate it was no one's turn so I walked up and merely stood near the ankle. Throughout this experiment the patient informed us very positively that the hands were those of the healer when he felt the strong movements resulting from this treatment. The absence of results he attributed to my own hands or the hands of no one. I wonder if, having read about the similar experiments in the book, you are able to predict what happened. In fact, the subject was quite clear in his mind when he thought the healer's hands were near his ankle and when they were not, but there was no correlation whatsoever between the movements of his ankle, with all the discomfort that followed, and the presence of the healer's hands. In fact, he sometimes experienced the strong movements when my own hands were there, sometimes when no one's hands were there and he often did not feel the movement when the healer's hands *were* there.

So what do we deduce from this? First, I asked the healer to step outside with me and I requested her on no account to tell the patient the result of the experiment. He was clearly benefiting greatly from the healing sessions and I did not want this to cease. However, I told her to tell him that she would give him 'absent healing' on every evening of the week at a specific time. He would then receive many times the benefit he was already having.

I should interject that the patient's doctors at a nearby famous London hospital had expressed no objection whatever to the healer carrying out her healing procedures upon him because, as she did not propose actually to touch him, there could clearly be no effect!

The healer agreed with me that her initial belief — that she was a channel for mysterious forces unknown to science and that it was up to scientists like myself to find out what they were — was somewhat astray. She agreed that suggestion appeared to be a very strong component in her valuable therapeutic system.

There are a number of loose ends which I have been unable to tie up regarding this method of healing. If, before the patient had met the healer, I had placed my hands on either side of his ankle, having no reputation for achieving cures, it is most unlikely that any movements of his limb would have occurred. So what started the whole process? I must confess to complete ignorance. Maybe a strong 'thought form' or belief structure had been created by the healer (see also the Philip Experiment on page 143) and this was picked up by the patient's unconscious (George). Normal suggestion would play a part, too. However, you would think that if a thought-form were the explanation then George would know whether, in the experiment, the healer was putting the hands there or not. You would not think it was necessary for the patient to see the healer as George's ESP would recognize her presence. We are clearly in deep water.

It was suggested to me by another investigator that the EEG pattern of the patient and of the healer tended to come into synchronism, to be similar, during a healing procedure. I do not know whether this is true or not and have not had an opportunity to find out.

Before leaving this particular healer there is a further comment which must be made. Certain other of her patients had spinal problems and when she put her hands near them they went into all sorts of peculiar contortions on the carpet, including arching their backs like cats. They continued to do this for some time and when the contortions began to ease off she put her hands near again and it all started up once more. The patients told me that nothing which they did in this automatic way (without any volition whatsoever from themselves) was in any way harmful but only therapeutic. It is very difficult to understand how this can be and it looks as though some part of the unconscious of the people present had the necessary medical and physiological knowledge. It really is very strange! I was reminded of statements I had read concerning the 'raising of kundalini' by appropriate yogic practices. The person in whom kundalini is rising is said to assume naturally certain 'asanas' or yogic postures and it could be that an Indian expert in kundalini yoga, observing this particular healer at work, might well comment that she is, in some way quite unknown to everybody present, manipulating kundalini. Again, it is a matter of making up your own mind. We have no Western-style evidence relevant to any of this.

RADIONICS

There is a method of diagnosis and healing which has been around for many years called radionics. This utilizes a 'black box' which carries a number of graduated knobs on the top and also a cylindrical container and a small area of rubber. There are various other features which are not relevant for the moment. The system was developed by Ruth Drown in the United States and by George de la Warr in the United Kingdom. The practitioner claims that, using this equipment, it is possible to diagnose and treat diseases. The patients do not need to be physically present provided the practitioner has a blood spot, sputum sample or a hair from the patient. The sample is placed in the small container and the practitioner asks – in the mind, not aloud – questions such as 'Is the disease below the waist?' He (or she) simultaneously strokes the rubber gently with his (or her) finger and if the answer is 'Yes' will feel some resistance to the motion – the finger tends to stick to the rubber. If the answer is 'No' then the practitioner asks, 'Is it above the diaphragm?' ... and so on. When the disease is finally pinpointed, say to the kidneys, the practitioner will ask in his mind, 'What is the percentage efficiency of the kidneys?' He will then revolve a dial which is graduated from 0 to 100 and consider that when his finger sticks to the rubber he knows the percentage efficiency of the organ of concern. In a similar way, by mental questions, he will determine what remedies and treatments to apply. These are usually homeopathic or herbal remedies as I know of no radionic practitioners who are orthodox allopathic physicians. Some of the radionic practitioners do not use the area of rubber in their diagnosis but hold a pendulum over that area. They have a code in which George tells them, via the automatic muscular system, whether the answer is YES or NO by the way the pendulum swings in either a circle or a straight line.

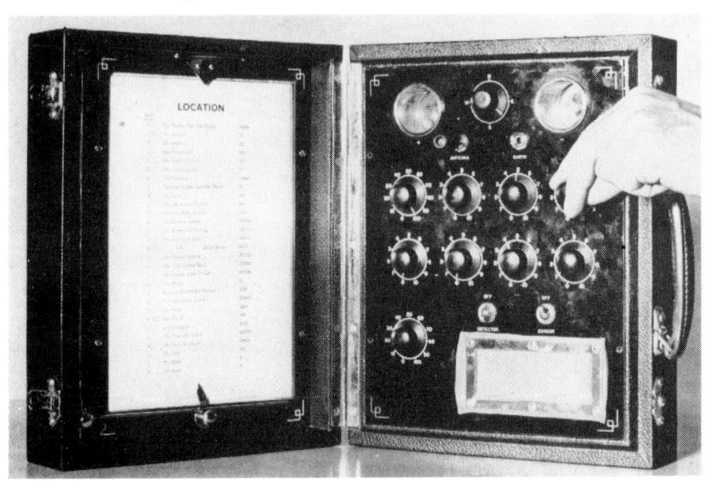

The radionics black box does not incorporate any known conventional technology. However, its 'scientific' appearance may help to win confidence and stimulate 'George's' co-operation.

When I tried this equipment a great many years ago I was quite surprised by the results. I put a sputum sample (on a piece of filter paper) into the container and rubbed the rubber area with my right forefinger, meanwhile asking the question 'Is this patient suffering from hay fever?' (I was earlier told by the practitioner that he was). My finger did indeed stick to the rubber – there was a distinct resistance to its sliding across. When I asked myself whether it was influenza my finger did not stick. I presume the explanation must be the production of moisture or perhaps a slight electric charge by my George: I have never had the opportunity to determine this.

It is important to note, of course, that George does not always give the right answer to unspoken questions, whether his replies or communications are made overt by a Ouija board, a hazel twig or by the rather different methods used by the radionic practitioner. This we need to test and see.

One particular method of treatment which radionic practitioners sometimes use is as follows. They erect what looks like a little aerial on the top of the box and place the box near a board carrying a number of blood-spot samples of their various patients, setting the dials on the box to what they refer to as the appropriate rate. They consider that the blood spots are in sympathetic vibration with the patients and that when the 'radiations' from the aerial treat the blood spots then the patient is being appropriately treated. (This is all perhaps somewhat of an oversimplification. I am however trying to give the principles which are claimed for this equipment and the general flavour of the method.)

A medical colleague and I carried out on the black box the following experiment. We asked a number of practitioners in various parts of the world if they thought they could, using their method of diagnosis, sort out a random collection of blood spots into male and female. It seemed to us that as practitioners claim to be able to establish quite small details over what exactly is wrong with a particular organ of the body, then it should be very easy for them to sort out blood spots into male and female, as some of the organs are of course quite different. All the practitioners were very happy to collaborate and stated that they would have no difficulty whatever in carrying out this task as they solved every day many diagnostic problems of a much more difficult kind.

So my medical colleague produced a number of blood spots from his patients and carefully labelled them with letters; the key (which stated whether they related to male or female patients) my colleague kept to himself. These blood spots were sent to another collaborator who placed them in other envelopes, putting his own labelling on the other envelopes. He therefore had no idea of the sex of the patient whose blood spot was within each envelope. So we sent the blood spots to each radionic practitioner in turn, in various parts of the world, and we assembled all their results. Perhaps you will not be surprised to hear that they had about fifty per cent of them right!

So the practitioners achieved only the results to be expected by chance in determining the sex of the patients whose blood spots they used – does this mean that radionics is of no use? Most certainly it does not. It could well be that the strong belief-structure of the practitioner plus the faith and belief of the patient in what is being carried out could indeed be of therapeutic value. The reader should remember that about seventy per cent of diseases presented to general practitioners are psychogenetic (produced by the mind) and would probably get better anyhow, without treatment.

What of the 'technology' of this system? The black box carries, as has been described, a set of graduated dials, the equipment inside being wired together in a certain way. The radionic practitioners – who are not scientists – use a language to describe the equipment, diagnosis and treatment, which sounds like the language of radio engineering. However, the wiring makes no sense to an electrical engineer like me and there appears to be no radiation, of any kind that I know about, from the aerial. Also, I know of no evidence to indicate that a blood spot is in some way in tune with its donor and that this can be used for diagnosis and treatment. This is not of course to say that there is no link between a blood spot and a donor: there probably is. But the link is perhaps a subtle psychic one and the diagnosis and treatment ought perhaps to be looked upon as a sort of technologically up-dated version of psychometry.

Another experiment was conducted some years ago when all the wiring was removed from the components of a black box and the panel replaced so that the practitioner was unable to see this. His success in using the box was unaltered.

There is one further point that must be mentioned. When the radiation 'treatment' is carried out with the box the dials are set to the appropriate rate for the treatment of the appropriate disease. Again, this does not seem to make normal scientific sense to me.

The treatment of patients by 'radiating their blood spots' has an equivalent application to crops. A photograph of a field is considered also to have links with the field itself and experiments have been done in which a photograph of a field has been 'radiated' by the aerial on a black box. Successful results were claimed by the radiators but when the botany department of a nearby university was brought in, the claims very rapidly evaporated.

PSYCHIC DIAGNOSIS AND TREATMENT

Let us now consider some rather more obviously psychic methods of diagnosis and treatment of patients. The well-known Edgar Cayce, who was a 'poltergeist child' (around whom poltergeist phenomena occurred) used to go into a trance and produce readings for patients, who did not have to be present. He could operate merely from the name and address. Cayce would refer in his trance to having found the patient and would then give a diagnosis and suggest a form of treatment. He sometimes found causes for diseases in 'past incarnations'. He was, according to all accounts, remarkably successful and his readings are preserved and are being carefully examined by his son and other researchers. Cayce also made predictions about future occurrences.

A most intriguing and interesting method of diagnosis was used by an Irish psychiatrist, Dr Connell, brother of the famous automatic-writing medium Geraldine Cummins. Connell would sometimes have patients referred to him who suffered unaccountable pains for which there was no physical cause and he tried a method which appeared to be very effective in certain cases. He asked his sister's 'control' to examine the past lives of certain patients to see whether there was some cause there for the unaccountable pains. The control did this and, in one or two remarkable cases, discovered such things as a patient's being killed in an earlier incarnation by a stab wound; the patient in the current incarnation having unaccountable pains at the point where the control said the stabbing had been inflicted. When the patient was given a description of what was stated to have occurred in the earlier incarnation the pain completely disappeared. (Those birthmarks which coincide with 'remembered' mortal wounds — in children who ostensibly recall earlier incarnations — are comparable; see page 128). Dr Connell said that he did not know whether reincarnation was true or not — he was perhaps not particularly interested — but he said that he was concerned to cure his patients and if this procedure did so then that was all he wanted.

A psychiatrist friend once tried age regression with me to see whether he could discover information about earlier incarnations to try to account for a pain which I have in my neck. Unfortunately although I can hypnotise *other* people, I do not appear to be a very good hypnosis subject and my depth of trance was not sufficient for the experiment to be successful. I propose to try other unorthodox methods of curing my unaccountable neck pain, all the orthodox methods having made no difference whatever over many years.

COLOUR THERAPY

Some healers treat their patients by relaxing them in the presence of light of various colours and consider this to have therapeutic effects. It is difficult to judge its effectiveness as it is often mixed with other treatments. So far as I know, no scientific work has been done to see if treatment by colour makes any difference other than that which can be explained by tranquillity, relaxation and belief.

VISUALIZATION

A most important and interesting method of treatment which has been widely applied to terminal cancer patients, has been deep relaxation combined with hypnosis, dietary control and visualization. The visualization part is of concern here. This procedure involves the patients having a clear understanding of exactly what is going wrong with their bodies and producing a clear visualization of what they want to happen – for example, imagining a shrinking of a tumour as a result of attacks by the white blood corpuscles. (Jack Schwarz's methods, referred to on page 110, of 'instructing George', are relevant.) The treatment also involves attempts by the therapists to give the patient a happy, comfortable and optimistic frame of mind.

Sadly, so far as I know, the beneficial results of this have been somewhat marginal but the work should clearly be continued. It may be that owing to lack of skill, the patients are unable to use their minds in this way, combined with appropriate suggestions to George, to produce the result. We do not know but when one considers the Philip Experiment (see page 143), in which physical energy was deployed by the use of a 'thought form', or the production of an apport (an object brought paranormally into the locked séance room – see page 50), it certainly seems worth while to deploy mental resources in an attempt to change the physical structure of the body. However, it may be that certain vital keys in the training of the patients are at present missing. One is reminded of the tremendous power over their bodies clearly evidenced by Swami Rama and Jack Schwarz (see pages 108-110). There is no doubt of the powerful effect which mind sometimes has on matter and a good deal more research is clearly needed. It is a great pity that governmental support for such research is not yet available.

PSYCHIC OPERATIONS

Many claims have come from the Philippines and Brazil concerning 'psychic operations'. In these a healer apparently removes material from the inside of a patient's body without leaving a scar. Large numbers of patients – usually from the local population but lately including a few people from the West – line up for treatment and the healer will treat a large number over several hours. There is no attempt at sterilization. Very often the healers will simply press their hands firmly into the soft tissues of a patient and matter which resembles blood then appears. Next the healer will take what appears to be blood-stained tissue from the body, puts it to one side and wipes off the blood, leaving no mark. The sceptical scientist will naturally think that the whole procedure is conjuring and any benefit which patients experience is the result of a strong suggestion plus a belief structure. This may not, however, be the whole truth. I cannot say too much about this subject because I have not examined psychic operations at first hand but I can say that scientific colleagues, in whom I have every confidence, have been to Brazil to examine the methods of the psychic healers and have filmed the operations. The filming does not enable one to say that sleight of hand is not being used and the results have been somewhat inconclusive. I have talked to educated Western people who have been to Brazil or to the Philippines for treatment and have returned stating that they felt very much better. However, this might be expected!

One or two psychic operations have been apparently performed in London and I have been told what occurred. In one particular case the healer drank large quantities of whisky during the operations and actually threw knives at the wall, narrowly missing the heads of some of the observers! He did, however, take out the eye of a patient from its socket and left it lying on the cheek while he did something behind it. This is of course quite possible in a normal operation and the only strange part is that an apparently medically-uneducated healer was able to do it without damage. Whether it was therapeutic or not I do not know – but it would certainly act as an exceedingly strong suggestion to the mind of the patient!

A psychic operation during which biological tissues are ostensibly removed from the body: is there more to this than conjuring and implied suggestion?

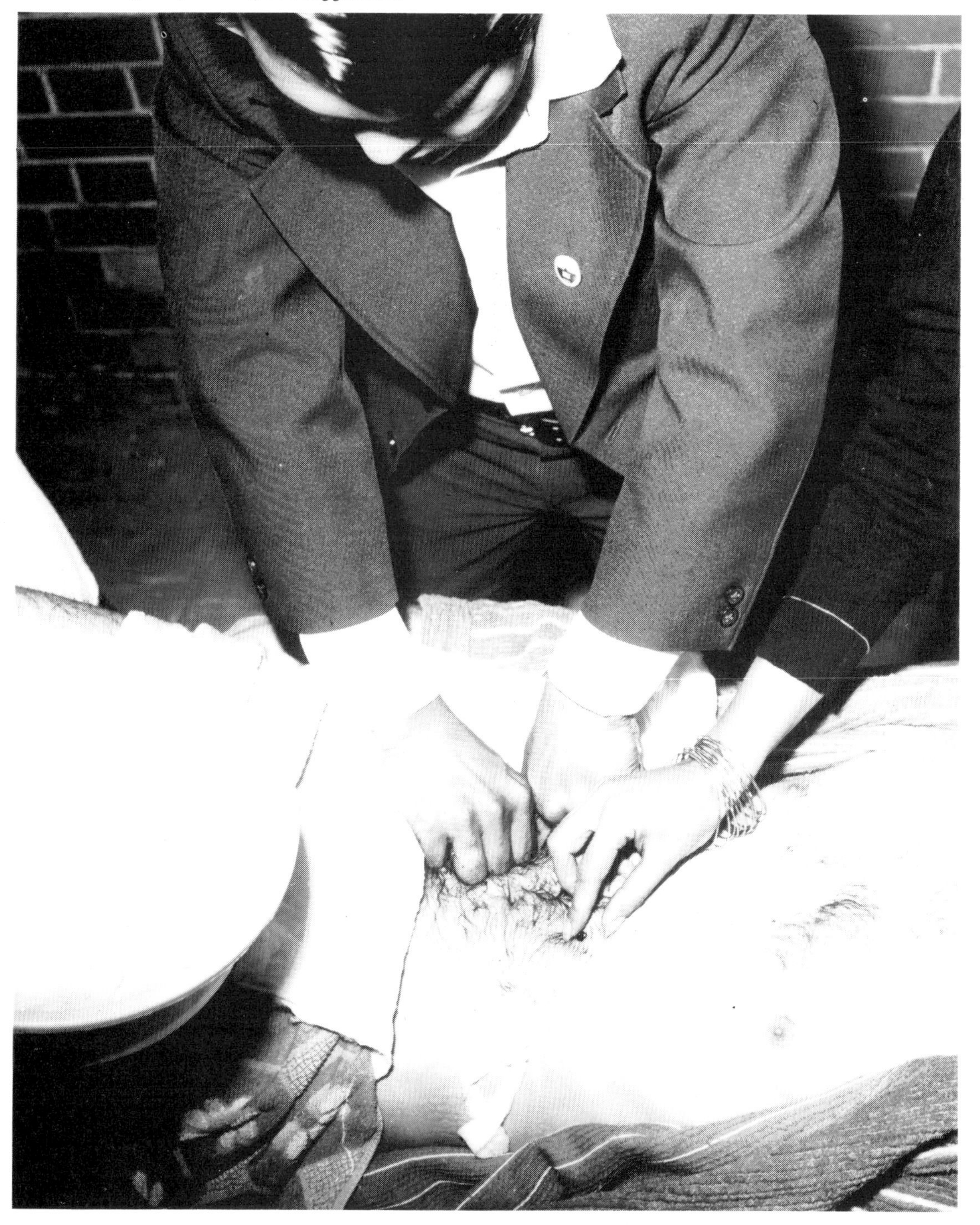

CONCLUSIONS

There is no hard and fast line between paranormal healing, complementary medicine and psychic functioning. It must be appreciated that many of the practices and procedures described in other chapters might be, and probably are, relevant to healing. Certainly if a patient practises regular meditation of a kind appropriate to them, then the resulting tranquillity of mind and the restoration of normality to bodily biochemical and physiological systems is likely to lead to a cessation of unpleasant symptoms. One must remember that the body has a natural healing power and, if you leave it alone, will often put itself right. It is primarily, I feel sure, the bad habit of our current lifestyle, an inadequate or unbalanced diet, lack of sleep and insufficient exercise, together with the ingestion of poisons and drugs of various kinds which perhaps lead to so much illness. This might be avoided by wiser living habits. Again, you must make up your own mind and make appropriate allowance for my own obviously biased views!

A final thought. I have within recent years talked to two distinguished specialists in cardiology. One of them told me that stress has nothing whatsoever to do with heart disease. He was investigating various surgical methods of treating such diseases. The other distinguished cardiologist told me that stress had everything to do with heart disease and his normal methods of treatment involved rest, sleep and relaxation — combined with advice to the patient on altering his or her living habits. So I was presented with two quite different views based upon a similar range of experience. Perhaps the most important thing one can say with regard to healing of any kind, is that a patient is a great deal more than an electrochemical machine and many people, sadly, do not know how to live properly. Whatever leads to good living habits, a tranquil mind and general happiness is also likely to lead to better health.

10
REINCARNATION

Belief in reincarnation — the idea that human beings have a series of lives on earth — is not common in the West but is exceedingly common in many other parts of the world. The idea is an integral part of the oldest religion in the world, Hinduism, and a similar idea was carried over into Buddhism, which sprang from Hinduism. However, the idea may not be all that unfamiliar to Westerners. I was surprised one Christmas when, following a general chat about parapsychology (in a hotel lounge after dinner), the subject of reincarnation was raised. I asked how many people present thought that there was some truth in the idea of reincarnation. Over half the people there said that they thought there was something in the idea and that it might be true.

DÉJÀ VU

Sometimes I am asked, 'Is the feeling that I have been here before evidence of reincarnation?' In other words, you are wandering around an unknown town during your vacation and you have a strange feeling of familiarity. You know what is round the next corner — and you are right. Is that not evidence of reincarnation? I think it rarely if ever is. So what is the explanation of this by no means uncommon feeling?

A likely explanation of the *déjà vu* experience is that you really had been there before — perhaps as a child — and had forgotten. Seeing some aspects of the place brings back to memory other aspects. Alternatively — and this is much more likely — you once turned over the pages of an illustrated travel book containing pictures of the place or watched a film or television production featuring the area. Suddenly seeing the vaguely familiar places again brings back recollections of others seen at the same time and subsequently forgotten. The memory storehouse is amazingly rich in information of which the conscious brain is oblivious. A third but much less likely possibility is that you had a forgotten precognitive dream, or some other psychic experience, giving you the information. However, this must be quite rare.

Neurologists know that some perceptions are 'tagged' as memories rather than perceptions. This is to do with the processing in the brain and is a common occurrence in temporal lobe epilepsy with discharging lesions of the right temporal lobe. The neuro-surgeon Wilder Penfield was able to produce déjà vu experiences in patients undergoing surgery of the brain for their epilepsy. He did this by electrically stimulating stuctures within the right temporal lobe. Some patients when so stimulated said, 'I know exactly what is going to happen. I've seen it all before'. (The right temporal lobe is often shown to be connected with 'psychic' experience.)

HYPNOSIS AND AGE REGRESSION

Much thought about reincarnation has been stimulated by radio and television programmes dealing with age regression under hypnosis. Reasonably good hypnosis subjects can be age regressed (that is, informed when in the trance that they are aged, say, four) and they will then behave exactly as they did (or give a fair imitation of this) at the chosen age, the appropriate memories apparently being present. If the age regression is continued back to birth and then to before birth and even further back, what is likely to occur? Usually descriptions of another life are given. However, the discerning observer will not immediately assume that this is evidence for reincarnation. George in our personal unconscious will very quickly realize that the hypnotist, by age regression to a period before birth, is looking for evidence of reincarnation. In the manner now familiar to us, George will look around in the memory store and see what he can find in order to dramatize pseudo memories of a past incarnation. It should be remembered that there is very good evidence that everything that has occurred during earlier times of this present life is stored somewhere in the unconscious and can in principle be brought out. Books about foreign travel, historical novels, and many other pieces of literature that we have read — or even glanced at — will be there for George to dramatize. There is a great deal of evidence that this does indeed occur. The first book (so far as I know) on this subject, *The Search for Bridey Murphy,* was written by Morey Bernstein in 1956, the subject being a girl in the United States who had never left the country. Southern Ireland was soon full of American reporters seeking for evidence that her ostensible memories of life in that country were in fact true.

There have been many similar books produced since then and they vary greatly in impressiveness. It must be appreciated, of course, that because George dramatizes memories and experiences to form pseudo memories of past incarnations this certainly does not mean that reincarnation never occurs. It may well do so. If indeed it does, and if memories are to be found somewhere in the unconscious (and the believers in reincarnation consider that they are) then George may well have available real memories of an earlier incarnation and these may be mixed up with the dramatized pseudo memories to form a mixed bag.

Some of the cases involving age regression under hypnosis are of a very different order of quality from others and certainly appear prima facie to provide evidence for reincarnation. The assessment of such evidence is exceedingly difficult. Memories which it is considered the subjects could not possibly have acquired during this life in the ordinary way may be produced under hypnosis but this does not at all prove that they did not at some time glance at a book or magazine containing the information, simply forgetting all about this insignificant incident afterwards. There is a well-known case of a girl who, when in trance, produced long strings of Hebrew verses, although she had no knowledge of Hebrew. However, it was later discovered that when she was a very small child she used to play outside the door of a rabbi who recited Hebrew verses out loud. This was before she was old enough to be conscious of what was occurring.

Some of the most persuasive evidence for reincarnation from regression under hypnosis is to be found in the Bloxham tapes described by Jeffrey Iverson in his book *More lives than One?* Arnall Bloxham dreamed of what he believed were scenes from a past life of his and then recognized a place in the Cotswolds about which he had dreamed, including a road and a castle. He was able to walk around the castle without a guide. At school he became interested in hypnotism; later he learned hypnotherapy and became distinguished in this field. He also acquired an interest in Eastern philosophy, including reincarnation. Bloxham cured a patient of an inordinate fear of death by helping him 'recollect a memory' of a previous life by hypnotic regression. The subject would not accept gentle suggestions from Bloxham that this earlier life was other than he remembered. Remembering the past in this way, Bloxham suggests, can help greatly with this life.

Bloxham was well aware that hypnotized subjects are likely to fantasize normally acquired but forgotten information and considers it important to demonstrate those memories that are distinct from the ones of this life. In any case the deadly dull nature of most of the lives he unearthed by regression is, he suggests, some sort of a guarantee of their genuineness.

Iverson includes in his book some of the more interesting and, in certain cases, historically verifiable, examples from the Bloxham tapes. One I particularly liked concerned a life of a Welsh subject, Jane Evans, who remembered a life lived as Rebecca, a Jewess. She ostensibly lived in the twelfth century in York and had a husband and two children. She describes, with graphic detail, strong emotion and a distinct period flavour, the persecutions and *pogroms* common during the reign of

Henry Plantagenet and his successor Richard Coeur de Lion. Especially interesting was their flight to take shelter in a church near the 'big copper gate' of York. She and one child hid in the crypt while her husband and son sought for food. The soldiers and mob burst in and her memories were then 'Dark ... dark' — evidently indicating her death.

The numerous details of the times, and the names quoted, were studied by a University of York expert in the history of that period. Barrie Dobson considered that the story was true to what is known of the events and the times. Much of the detail was accurate and some disputed parts could have been true; these included scholarly aspects which only professional historians might have been expected to know. Her 'copper gate' was confusing. There is a street called Coppergate but apparently it was possibly a street of joiners; the word may have been derived from the Scandinavian Koppari meaning joiners. The church described was quite definitely St Mary's Castlegate, but nothing of any crypt was known. Six months afterwards Iverson visited St Mary's and, while the building was being converted into a museum, a crypt was found by a workman, who described having seen arches and vaults. It is important to note that Rebecca's story is not at all the history-book version. Important facts of the history of the time are omitted and the whole gives an impression of an eye-witness rather than of someone fantasizing material read earlier.

CHILD MEMORIES

So what is really good evidence for reincarnation? Probably the best evidence for reincarnation is the memories some very young children profess of another life; memories which are later confirmed as accurate, even though the young children could not possibly have known the relevant facts in any normal way. There are thousands of cases which approach in quality this standard. Dr Ian Stevenson, Professor of Psychiatry at the University of Virginia, has spent a considerable portion of each year travelling the world to discover cases of children who remember what appear to be the facts of an earlier life lived by them. With children one can usually be fairly certain whether the apparent memories could have come from normally acquired information. If they are having spontaneous memories of previous lives the children usually begin to talk about them as soon as they can speak — at the age of two or three years — supplementing their speech with gestures. At this stage the parents can usually say who has been in contact with the child. The numbers of memories vary over a wide range. Usually the quantity and clarity of the memories increase until a certain age, perhaps five or six, and then the child starts to forget. Sometimes some of the memories are able to be recalled for a considerable period. Memories are not the only factors remarkable in cases of this kind. Sometimes a child shows unexpected behaviour, unusual for the family in which the child was born but appropriate to his or her claimed earlier life. And this behaviour is often found to correspond with what others say concerning the behaviour of the deceased person — 'the previous personality'. The unusual behaviour may involve particular phobias concerning, say, weapons of a particular kind or strong likings and dislikings for various objects. Often the children show a certain dignity which appears inappropriate considering their young age and sometimes they may be patronizing towards children much older than themselves. The children, in fact, feel themselves to be adult. They often have birthmarks or deformities corresponding to a fatal wound on the body of the claimed previous personality. For example, one subject showed a birthmark which had the shape of the blade of a particular spear used in the district of the remembered previous life. Or there may be a birthmark at the front of the body and a corresponding one at the back, with a memory of death by shooting. In one case I recall, the birthmark resembled the healed entry point of a bullet.

Sometimes the expectant mother (or a friend or relative) has a dream in which the deceased person indicates his or her imminent arrival as the new baby. The arrival may be stated directly or — in the characteristic way of dreams — symbolically. Stevenson has discovered that the interval between death and presumed rebirth varies from nine to forty-eight months, particular periods tending to recur for particular cultural groups.

SHANTI DEVI

Probably the best known and reported case of ostensible reincarnation is that of Shanti Devi, who lived in Delhi (she was born in 1926) and who began to describe details of a former life lived in a town called Muttra some eighty miles or so away. She had these memories from about the age of three and stated that her former name was Lugdi, that she had been born in 1902 into the Choban caste and had married a certain Kedar Nath Chaubey, a cloth merchant. She said that she had died ten days after the birth of a son.

Eventually, when she was nine, the family wrote to enquire whether a Kedar Nath Chaubey did indeed live in Muttra. He himself answered the letter, confirming the girl's statements. Then he sent a relative to Shanti Devi's home and afterwards came himself unannounced. She at once identified both of them. In 1936, after it was quite certain that Shanti Devi had never left Delhi, a committee was formed to go with her to Muttra and to observe. At the railway station in Muttra she picked out another relative of Kedar Nath Chaubey from a large crowd. She was then put in a carriage and the driver asked to allow her to direct him. He drove under her instructions to the home of Kedar Nath Chaubey (painted a different colour from that of earlier years, which she remembered). Near the house an old gentleman appeared and she at once identified him as her earlier father-in-law. She correctly answered, within the house, several questions concerning such matters as rooms and cupboards. Later she visited the home of those she claimed were her previous parents and again picked them out of a crowd of people, calling them by name. An impressive feature of this case was her use of the local Muttra idioms of speech.

This is a brief account of the Shanti Devi case. However, it is important to note that there is no mention in the records of any incorrect statements made by her and she made perhaps twenty or thirty verifiable statements.

THE CASE OF JASBIR

The most interesting and unusual case described by Stevenson occurred in Northern India. Jasbir, a three and a half year-old son of Sri Girdhari Lal Jat of Rasulpur in Uttar Pradesh, was thought to have died of smallpox. The father asked his brother and others of the village to assist him in burying his son. It was late and the burial was postponed until the morning. Later the father happened to notice the body of his son stirring and he then revived. After some days passed he was able to speak, expressing himself clearly a few weeks later. However, he showed a most remarkable change of behaviour, now stating that he was the son of Shankar of Vehedi village and wished to go there. He would not eat with the Jats, as he said that he belonged to a higher caste and was a Brahmin. A local Brahmin lady kindly cooked for him in the appropriate manner or he would certainly have died. Gradually they weaned him away from this habit.

The boy, Jasbir, began to give more detail of his earlier life and death, stating that he had been given poisoned sweets by a debtor of his and had consequently fallen off the chariot during his wedding procession, dying from a head injury.

The details of the death, and other items stated by Jasbir, corresponded to the details of the life and death of a young man of twenty-two, Sobha Ram, son of Sri Shankar Lal Tyagi of Vehedi. He had died in exactly the manner described although the family knew nothing of any poisoning or debt. (They were afterwards suspicious.) These details were reported to the family as a result of a lady visiting Jasbir's village and being recognized by him as his aunt. Later other members of the family in Vehedi visited Rasulpur and met Jasbir. Jasbir recognized a number of members of the family and correctly referred to their relationships. Jasbir was later taken to Vehedi, put down near the railway station and asked to lead the way to Tyagi property. He did this with no difficulty. Remaining some days in the village he demonstrated his detailed knowledge of the family, returning home reluctantly.

Professor Stevenson visited the two villages and interviewed many witnesses of the case, returning several years later with new interpreters and interviewing most of the previous witnesses and some new ones. He refers to the extremely unlikely possibility that the child could have obtained the information he had in any normal way. A rather elaborate theory, involving telepathic links — extended well beyond the limits for which there is evidence — would be needed to explain this in that way. The particularly interesting point about this case is the remarkable change which took place when the child was three and a half years old. The case therefore differs from others in this respect.

A point of particular interest to me related to Jasbir's answer when Stevenson asked him if he knew what happened to the mind or personality that occupied the body of Jasbir between the death of Sobha Ram and before that body had seemingly been taken over by the mind of Sobha Ram. He replied that after death he (Sobha Ram) met a holy man (a sadhu) who advised him to take over the body of Jasbir who had ostensibly died. Stevenson has, despite enquiries, been unable to trace a child who claimed that in a previous life he was Jasbir of the village of Rasalpur who died of smallpox at the age of three. Jasbir told Stevenson that in dreams he sometimes saw again the discarnate holy man, who had advised him to 'take cover' in the body of Jasbir. He also referred to certain correct predictions of future events in his life given to him by the holy man in dreams.

THE POSSIBILITIES

Cases of the reincarnation type are found much more frequently in those parts of the world where there is a widespread belief in reincarnation. However, there have been quite a few cases reported and investigated in cultures where most people are uniformed about reincarnation or even opposed to a belief in it. There have been a number in the United States, Canada and the United Kingdom. It seems very likely that more cases occur in the West than are heard of but that these are suppressed by the parents of the children concerned.

It is important to emphasize the enormous care taken by Ian Stevenson in obtaining independent verifications of the statements of the child concerned — the child's parents' word alone being regarded as quite insufficient. There appear to have been very few fraudulent cases. Usually there is no motive for fraudulence as, in most parts of the world, money cannot be made out of such matters. Where the two families concerned have met, and very often this occurs before an investigator arrives and begins to produce records, the families may, quite unwittingly, mingle their memories of what the child said with what they learn or remember about the previous personality and may then attribute some of this to the child. This may apply, at least partially, to a few cases but is very unlikely in the many cases which are rich in remembered details. Another possibility is that the child heard a lot about the previous personality in a normal way, forgot all about it, and then it was dramatized (by George) to appear like normal memories. (This phenomenon is called cryptomnesia.) However, most of the families live in villages or towns separated by many miles and had no acquaintance with each other prior to the development of the case, so normally this hypothesis is very unlikely. Also, as the children often start to speak of their memories at between two and three years of age, it seems most unlikely that detailed information could have been communicated to them without this being known and remembered by the parents.

Another possibility which has to be considered is that the child obtained the information by ESP and then the unconscious mental machinery we have called George created from it the previous 'personality'. This could not explain a commmon feature of such cases, namely that the memories contain no knowledge of changes in topographical features of the old environment (such as the colour of a house or the construction of buildings) which would have occurred since the death of the ostensible previous personality. In addition, some subjects have shown remarkable skills of one kind or another and it is difficult to see how this could have been obtained by ESP. Identification with the previous personality, together with attitudes and strong emotions are also difficult to imagine being transmitted by ESP. The situation is by no means a simple recital of information.

A very important point which must be mentioned is the occurrence of cases of this kind in cultures effectively isolated from each other for many centuries. This suggests a widespread human experience. Another point is that the children who have these ostensible memories of an earlier life generally grow up as perfectly normal children and forget these memories. If any reader has, or knows of, a child who has memories of this kind, the author would be glad to hear of it, but no concern as to the child's health need be felt.

In the large collection made by Stevenson it is interesting to note how small is the number of cases in which the life is remembered as a person of the opposite sex. One researcher reported an unusually large incidence of fifty per cent sex-change cases in a group of forty-four cases discovered amongst Indians in Canada; the Stevenson (University of Virginia) collection drawn from Burma and Thailand (a study of 250 cases) showed about twenty per cent sex changes. In all other cultures, however, the incidence is about five per cent.

Stevenson points out — and it should certainly be considered — that it is quite possible that the belief held by a person before he dies (probably reflecting the general beliefs of his or her culture) may well influence what occurs. This is an old Buddhist idea and certainly seems eminently sensible to me.

The evidence that at least certain people appear to have reincarnated is very good indeed. Therefore this may be worth considering as a possible explanation for the way certain children display remarkable abilities or unexpected behaviour which cannot be accounted for in other ways. For example, the remarkable propensity for playing the piano exhibited by the young Mozart has often been suggested as evidence of reincarnation. Reincarnation might also be considered as a possible reason for excessive effeminacy in boys and masculinity in girls: they may have had an immediately preceding incarnation in the other sex. Perhaps homosexuality is a result of such a change. It may be a relevant factor. But we are just speculating ... who knows?

A PERSONAL VIEW

So far I have not actually stated my own views about reincarnation. Needless to say, I have given a great deal of thought to the matter over many years. My ideas have changed very much over the time that I have studied this subject. I remember being greatly affected by the ideas to be found in the Veda (Hindu scriptures) and feeling that reincarnation linked to the Hindu idea of karma (cause and effect: 'as you sow you reap') provided an almost perfect and certainly very logical philosophy of life, leading, if accepted, to fortitude and peace. If the troubles which almost all of us have in our lives are the results of causes we ourselves have set going in the past, then a philosophical acceptance should be a result. Moreover, as the idea of karma implies that it is up to the individual how he accepts his karma and reacts to it, and this decides his future, then surely the future should be one of hope. I have often listened to Christians in the West struggling for explanations of the manifest inequalities of life and have considered how little problem these give to a Hindu with his belief in the twin doctrines of reincarnation and karma.

There is evidence indicating that a fair proportion of Christians in the early centuries of the Christian era did in fact believe in reincarnation and it is implied in various parts of the Christian scriptures. However, it was officially rejected, I believe, in the sixth century AD, as a result of a vote by church leaders, probably none of whom had any firsthand knowledge of the evidence! Certainly the great Christian mystic Origen believed in reincarnation. It is interesting that the orthodox members of the religion of Islam also do not officially believe in reincarnation even though it is similarly implied — in fact stated — in their own scriptures. Numerous Western writers and others have or had a belief in reincarnation, and they range from Napoleon and

Salvador Dali to W.B. Yeats, Goethe, Conan Doyle, Tolstoy, Edgar Allan Poe and Louisa Alcott. The psychologists William James and C.G. Jung and many engineers and scientists — like Henry Ford, Thomas Edison and Sir Oliver Lodge — have taken the subject seriously.

However, all these beliefs do not amount to scientific evidence, which is on the sparse side. The evidence, such as it is, from 'communications' through mediums does not seem to indicate that, if things are as they seem, the 'communicators' know any more about it for certain than we do here. If reincarnation is generally true then it takes place from a level of consciousness inaccessible through the activities of the average psychic.

An idea I read many years ago which appears to be acceptable to a number of wise people for whom I have a great deal of respect, is that a human being is somehow divided into two — a personality with emotions and mind, and an 'individuality'. The individuality functions at a much higher spiritual level and pays not too much attention to the personality until it seems to be showing signs of taking the right path — the right path being the spiritual path of growing unselfishness and dispassion. 'Individuality' might be an appropriate word for the part of a human being referred to by Saint Paul as 'the Christ within'. If this idea of the structure of a human being — and it is by no means a new one — has truth in it then reincarnation would involve the individuality investing a series of emanations from itself into personalities. In other words, a succession of different incarnations would each give certain experiences and lead to particular 'spiritual fruits' which would be used to help the development of the individuality. A new incarnation would involve a quite new personality with a new physical body, emotions and mind, and the memories of all the previous personalities which had been into incarnation would then be held only by the individuality. There would be no memory in the mind or memory store of each separate personality. It would be necessary, on this theory, through spiritual development, to raise the consciousness to the level of the individuality before any knowledge of past incarnations could be gained.

It is interesting that Edgar Cayce, the exceedingly competent psychic who is so well known for his psychic readings (relating to medical treatment and healing) and for his life readings (relating to the future), stated that in the unconscious were to be found a record of all the past incarnations of an individual. He received this idea during one of his trance occasions — to his shock and dismay as he was a believing Christian and the idea of reincarnation was anathema to him. However, he grew used to the idea and found much of value in his readings involving the reincarnations of various of his investigated subjects. (See page 122 for a closer consideration of Cayce's psychic information.)

Individuality: this works on a higher spiritual level, not paying too much attention to everyday matters.

Personality: this is involved with immediate experience, focussing on emotions here and now. We can add to the individuality (or 'spiritual bank') only through acts of unselfishness.

It might well be suggested as evidence against the doctrine, that ordinary human beings have no memory of past incarnations. However, in the case of a young person who died a violent death, it is arguable that possibly a reincarnation could take place which still uses the same mind. This mind

One theory suggests that we accumulate experience through living different lives as different personalities and that this contributes to the overall self or individual. Although we have no memory of previous lives the cup of knowledge and understanding is gradually added to.

would then contain the memories through which it had already passed. However, being reintegrated with a new brain, the memories might be expected to be difficult to bring through and perhaps would fade with time.

If I had to say what I thought was probably the most feasible suggestion I would choose to support this last idea. I have a feeling (certainly far short of proof) that human beings do have a series of lives on earth and that the fruits are gathered into some sort of individual spiritual store house. Perhaps one day in the far distant future this store house becomes accessible to us. In the meantime, the experience is perhaps a major influence in our conscience — what we can and cannot allow ourselves to do. Those children who remember an earlier incarnation are bringing the mind of that incarnation, with its memories, back into association with a new physical brain. It is an exciting and far-reaching idea.

Is there anything practical which we should do, or are there ways in which we can explore further the ideas and evidence of this chapter? If we are especially interested in reincarnation a reading of the Veda and Upanishads might be stimulating. (The Bible also contains references to the belief.) However, if we know children who have ostensible memories of another life we can show sympathetic interest and refrain, as so many do, from suggesting that the children are making up stories. If there seems any practicable way of checking the 'remembered' facts we can try to do that. If reincarnation is true it would seem to be very important for us to know it. It could make a big difference to our outlook on life.

11
ASSESSING CURRENT EXPERIMENTS

APPROACHING THE RESEARCH

It is my hope that the result of reading this book so far will be to convince the reader that parapsychological phenomena really do occur and we cannot just dismiss them as being impossible. The purpose of this chapter is to consider how it is possible to use the normal and very successful methods of science to produce a better 'understanding' of the phenomena. The word understanding is in inverted commas because, as has been suggested earlier, science is the process of building mental models representing our experiences and if those models fit and enable future experiences to be predicted they become the basis of our understanding. The point is that we cannot understand something of a fundamental kind regarding objects 'out there' in the way we can understand a geometrical theorem. In understanding a model representing our experience we are still one step removed from understanding how we are having such an experience; we understand only the mental model *which represents it.*

Now how do we use the ordinary methods of science in this way? First, we attempt to fit the new experiences we are having — which here we are calling parapsychological — into the mental models we already have. In other words, we try to use our ordinary Western scientific models to explain the new facts. If we cannot, and if we are quite certain that the new phenomena really do occur, then — I hope — we attempt to model them in the simplest possible way. Good science is always clear, simple and unequivocal.

It will be remembered from an earlier chapter that many psychics have the idea that the astral body is made of subtle material interpenetrating the physical. This theory was tested by covering up the physical body and observing whether psychics could determine its position in space by their psychic observation of the astral body. It was found that they could not do this. The psychic observation of the astral body cannot then be modelled in the same way as we model physical objects in space all around us to represent certain experiences in our minds (see page 97 and 99).

A similar example but with a different result relates to the psychic observation of the etheric double, which is supposed to be, again, some semi-physical material interpenetrating the physical body, projecting a few millimetres all round, and being used to carry certain 'vital forces of life', called *prana, c'hi* or *Qi* and *kundalini* in the East. One method which some books advocated for seeing the etheric double was as follows. The fingers of the two hands were to be placed a few millimetres apart in a dim light — the red glow of the dying embers of the fire was recommended — and the experimenter was to use his or her will-power to cause the etheric matter to flow between opposite fingers. In most cases it should then be possible to see the etheric matter.

I tried this and was indeed able to 'see' the etheric material. There is however a normal explanation for this effect. If you put the fingers near one another in the way described, then the eye, in examining the two hands, moves quickly from left to right in the way we would always use for looking at two objects close together. The effect of this is to leave faint after-images on the retina which appear to fill in the space between the fingers and look like greyish material. So the phenomena can be explained by using no more than our present models. A similar explanation was given, again in terms of ordinary scientific models, with reference to the use of the Kilner screens to see the etheric or astral body (the two terms are sometimes used vaguely and interchangeably). Again, the phenomena were compatible with the ordinary understanding we have of vision and light (see page 100).

If strange new psychic explanations for perceptions are being put forward then it seems to be essential to consult with scientists who are quite familiar with the current scientific models representing similar experiences. (This assumes that the investigator is not already a scientist who has appropriate training and experience.)

I remember some years ago experimenters tried listing a number of substances in order of their effectiveness in stopping the flow of etheric material. This was done as a result of a single psychic examining the flow of etheric material when the substances were placed in its path. The experimenters had no difficulty in placing them in that order as they worked with only one psychic. Unfortunately, this experiment is not sufficiently fundamental: George in the unconscious will certainly alter the apparent density of a greyish image if different substances are imagined to be inhibiting the flow of this material but if the greyish image can be explained by the ordinary and well understood laws of optics and vision then there is nothing objective about the order in which the substances are placed. The whole experiment results in a piece of useless fantasy, as a different psychic and different experiments would probably have put the substances in a quite different order.

A similar example of an experiment with a psychic which was insufficiently fundamental went as follows. The psychic was describing, using the magnifying power of the *ajna chakram* — in other words, his psychic faculties — the passage of an electric current through a piece of carbon. The descriptions were carefully examined to see how they fitted in with the normal scientific ideas of electrons flowing through material and an electromagnetic field surrounding it. However, when the psychic did not know whether the current was actually on or off, as a result of using a randomizer machine, then his observations became exceedingly vague and he began to talk of being in 'the eternal now'. It is exceedingly easy to be deceived and produce fantasy when the mind of a psychic is involved in this sort of way.

The important point which must be made about all this is that psychics have facts of experience which do not fit current scientific models. It is essential, in studying these experiences, to have scientists present who thoroughly understand all the present scientific models. This is no place for an amateur! In examining the claims of a psychic — and Uri Geller is a very good example of someone who has been examined — it is vitally important to include in the scientific team who plan and assess the experiments, scientists representing all the relevant disciplines. They will then be able to say whether anything is occurring which they cannot explain using the current models. In the case of an entertainer like Geller then expert magicians are also vital members of such a team and, if the examination is of paranormal metal bending, then a metallurgist is clearly another essential member. The team which I set up on behalf of the Society for Psychical Research to examine Uri Geller (and which unfortunately never was able to make the arrangements to do so) included magicians, physicists, engineers, a psychologist, a philosopher, and a neurologist, with representatives of other disciplines available — as necessary. A scientific examination of the magnitude which was planned necessarily would have had to take place within a university or hospital.

Let us then come on to the way in which the psychical research must be carried out when the phenomena (the experiences) cannot be explained in terms of the ordinary models of science but when they are reasonably reliable and adequately repeatable. (The reason I have added these qualifications will become clear in a moment.)

Psychical research has to be carried out and consolidated in the same way as any other science but there is a great deal more to it than there is in many other branches of science. It is necessary to consider as vital components factors that can usually be safely neglected in other subjects. Those factors which are of great importance in psychical research are to do with the minds of the people involved. This idea is not new to quantum physicists but it is certainly unfamiliar and disturbing to many other types of scientist. The problem of good psychical research is that it is difficult to examine, measure and record the mental state of everyone connected with an experiment in the way we should like. However, we have to do the best we can. All that is possible, in general, is to measure electrical brain rhythms (EEG activity) — these are the physiological correlates of mental states — and carry out psychological tests and inventories. Important work has been done by psychologists in devising appropriate psychological inventories which are of value in this subject.

Some scientists without any experience of psychical research sometimes pontificate on the lack of repeatability of much of the work done by parapsychologists. They say — and this is well known and usually true — that if the protocol of an experiment in ordinary science is set out exactly, then any other scientists should be able to repeat that experiment and achieve exactly the same result. This cannot be done normally with parapsychology because of the difficulty of establishing the mental state of the people involved in the experiment. In other words, ordinary scientists who make such statements are doing so, usually quite unwittingly, on the basis of a very naïve realism, that is, they assume that the universe is out there, independent of themselves and that they can observe it without affecting it. They also assume that the result of their action, or anyone else's similar action, will always be the same. That is nearly always true for most ordinary science practice — if you ignore the odd occasions when it is not! However (as I have restated *ad nauseam*) even ordinary science is the building of mental models representing experiences, and any psychologist will point out that the mind itself adds to the sensory inputs to construct most of what is being 'observed'.

A REVIEW OF REPEATED EXPERIMENTS

During the course of their experiments, some phenomena become well known to all parapyschologists because they have been verified so many times in different laboratories.

One of the best known phenomena of parapsychology is the so-called sheep/goat effect, originally named by Gertrude Schmeidler. This is the fact, many times repeated and established, that belief in the possibility being tested helps to produce results, while disbelief hinders. Subjects involved in ESP experiments are much more likely to produce successful results (results that are better than chance expectation) if they have a view that ESP could well be possible. If they have a very firm view that such phenomena are quite impossible and conflict with the laws of nature then, carrying out exactly the same experiments with exactly the same experimenters as the others, they are more likely to achieve results below the chance expectation. It is important to realize that results greatly below chance expectation are just as remarkable as results greatly above chance expectation: both appear to show that there is ESP knowledge present and, in the case of the low scorers (the goats) the subject's George is deliberately suppressing the results to support the view that it is impossible. Unfortunately — we have already pointed out how a-logical George is — doing this is evidence for ESP which is just as good as high scores. This sheep/goat phenomenon has been verified many times in numerous laboratories and, in my mind, there is very little doubt about its existence.

Another very important effect which is also well-known to parapsychologists is the so-called experimenter effect. I am also quite certain that some experimenters, carrying out identical experiments with the same subject, are much more likely to achieve successful results than are other experimenters. The experimenters whom one might expect to achieve success rather more often than the others are said to be catalysts; the others are said to be inhibitors. A number of my friends, good parapsychologists all, can most certainly be divided into catalysts and inhibitors. I think, from the many interesting experiences I have had in connnection with parapsychology, that I am probably a catalyst. I am not a psychologist and therefore not competent to devise psychological tests to divide experimenters into catalysts and inhibitors, but I have certainly noticed some personal characteristics which do appear to correlate with them.

The successful experimenter, I have noticed, is very likely to be a warm, friendly, smiling individual who gives subjects an optimistic atmosphere and

ARE YOU A SHEEP OR A GOAT?

1 Do you believe it is possible to communicate telepathically with another person?
a.No
b.Yes
c.Uncertain

2 If a friend tells you he has visited a faith healer and been cured of an acute medical problem, would you believe this was due to:
a.A natural improvement in your friend's condition which would have occurred anyway — with or without the healing session?
b.The powers of the healer?
c.Your friend's own belief in the cure?

3 Do you think that metal bending (as practised by Uri Geller) is:
a.Impossible, or the result of a clever conjuring trick?
b.A convincing paranormal phenomenon?
c.Something you might believe in but only if you saw it happen first hand?

4 The telephone rings and you know who is calling you before hearing their voice. Do you attribute this to:
a.Coincidence?
b.Extra-sensory perception?
c.Logical powers of deduction?

5 Do you believe that dowsing with a twig or divining rod in order to find water or hidden objects is:
a.Frankly impossible?
b.A special psychic skill which is possessed by a few gifted people?
c.An old-fashioned proven method?

6 Do you think that ghosts are:
a.Non-existent?
b.The spiritual forms of people who have died?
c.Apparitions produced by the 'imagination' of the witness concerned, but which can look very real?

7 You experience the sudden overwhelming sensation of having been in a particular place or situation before. Do you consider this to be the result of:
a.A trick of the brain?
b.Information percolating through from another life, or perhaps the ability to see into the future?
c.Having seen the place or incident before, maybe on television or in a book?

8 If a friend asks you to accompany them to a séance, would you go:
a.Expecting an evening of sensationalism and fakery?
b.Looking forward to the opportunity of witnessing some psychic phenomena?
c.With an open mind, prepared to evaluate carefully any unusual happenings?

9 If asked to predict a sequel of random numbers or symbols would you expect to score:
a.Badly?
b.Above average?
c.An average result?

How did you score? A high proportion of a *answers means that you are quite definitely a Goat. Sheep would have a fair number of* b *and perhaps a few* c *answers. Those with a majority of* c *answers are 'cross-breeds' — yet to be convinced but ready to examine the evidence.*

has with him or her an air of quiet confidence; the inhibitor is much more likely to be rather cold and analytical, smiling rather more rarely, and adopting a more clinical approach to an experiment.

Perhaps as a result of their personality characteristics, the inhibitors are far more pessimistic concerning the possibility of the results they may be seeking and this pessimism communicates itself, one way or another, to the subjects.

It was the great pioneer J.B. Rhine who defined a good psi experimenter as not simply one who rules out all normal cues to the target and administers a randomized set of targets to the subject, but rather as one who can succeed in 'liberating the psi function' under adequate test conditions. In other words a good experimenter – a catalyst – should be able to provide the psychological conditions under which the psychic faculties operate. Even though those conditions cannot now be described, most parapsychologists certainly agree that they exist and it must be very disheartening for keen experimenters if they go to a great deal of trouble to try to carry out some good experiment in parapsychology only to discover that they have a personality type which is unlikely to succeed.

I have already mentioned that an experiment is a Gestalt; that is, everybody concerned with it, together with the surroundings and the protocol, make a unified whole. It was Gardner Murphy who suggested (it is not necessarily true) that there is no such thing as a gifted ESP subject but that whether or not the subject scores well in a test depends on the person, or persons, who do the testing and the nature of the experimental conditions. There are outstanding experimenters and others who rarely have success. Most parapsychologists could say whom they admire most and consider to be good at this but, as we have already said, it is impossible to be too precise.

The successful parapsychological experiments with which I have been associated I remember particularly as generating a tremendous amount of interest and enthusiasm from everyone concerned. That excitement of being associated with something definitely occurring and very unusual is perhaps an important factor in maintaining the success rate. Some experimenters are simply unable to develop such a pitch of interest and enthusiasm in their subjects, whether or not they appear to be achieving initial success.

LABORATORY EFFECTS

I remember well some years ago inviting for controlled tests two children who experienced psychic phenomena in their house following the Uri Geller broadcasts (see page 59).

I invited them to London with their mother (they were a girl aged eleven – and a boy of thirteen). We spent the mornings doing experiments and afternoons entertaining the children, who had never been to London before. They achieved quite remarkable success in several experiments in their own home but, to cut a long story short, they had no success of any kind in the University laboratory. One of the children was able,at home,to bend paranormally a teaspoon (that I had bought myself) by gentle rubbing between finger and thumb and later was also able to produce a sketch remarkably like a cartoon in a newspaper which I had brought in that morning (that particular day's newspaper) and which the family did not take; no one present had actually seen the cartoon. The girl drew the sketch on a piece of paper placed on top of the folded newspaper, which was not unfolded and examined until after the experiment.

When the children came to London and entered a university laboratory (for the first time) they were surrounded by mysterious equipment and people of a kind they had not earlier met. They suffered what I have since described as laboratory inhibition, that is, all their psychic faculties were stultified by the laboratory. Future experiments of that kind should of course be designed so that they can be carried out in the subject's own home or somewhere they feel safe, secure and comfortable.

There is another phenomenon which must be mentioned in connection with laboratories. Often, when psychic occurrences are being studied, one uses delicate measuring equipment in order to detect small effects. The experimenter uses the laboratory transducers, amplifiers and other equipment to make overt small changes in some physical parameters. The result of this is, not infrequently, to produce results which appear to be positive whereas actually they are perfectly normal and the result of using the equipment in an unusual way. A case I remember particularly related to some tests concerning 'thought and emotion transfer to a plant'. Sir Alister Hardy, the well-known biologist, directed to me a researcher who claimed that he was observing paranormal electrical activity produced by a plant when he had the intention of damaging it.

Professor Hardy requested my co-operation in providing laboratory space and measuring equipment, with electrical expertise, so that the truth or otherwise of this claim could be determined.

After a few days' work in the laboratory the investigator called me down to demonstrate to me that he was receiving the expected results. He had attached a microvoltmeter to two electrodes, clipped to a plant (which I had obtained for him from a well-known botanical garden in London). He indicated to me that he was thinking of burning the leaf of the plant to which he had attached the electrodes and pointed out to me that the microvoltmeter was 'going wild'. However, it appeared to me to be showing considerable movements when we were chatting together and he was not particularly thinking of damaging the plant at all. So I tried an experiment.

I rubbed my foot firmly across the floor several times and the voltmeter really did go wild. That, I pointed out to the experimenter, was the result of electrostatic charge. I suggested that he take the plant and the instrument to another laboratory where he could place it within an electromagnetically screened room and exclude the effect I had demonstrated. He did this, and sadly left the University a day or two later. All his paranormal claims had evaporated!

Another experiment I remember illustrates something else which not infrequently occurs in parapsychology, that is, a slight extension of our normal knowledge of a subject that has nothing to do with parapsychology. One such experiment related to a healer who claimed that he could protect blood cells, which were placed in dilute saline in a test tube, from disintegrating and releasing their haemoglobin. This disintegration is called haemolysis. The healer informed us that he had achieved success in the United States with another experimenter. So several of the experimenters visited the University Health Centre and there the Nursing Sister extracted from each of us blood samples. For each test with the healer, a measured quantity of blood was put into a measured quantity of the damaging saline liquid in a test tube. Then the subject for a measured time attempted to protect in turn each sample of the red blood cells. Other specimens, the controls, were treated in a similar way except that the healer did nothing. The degree of haemolysis of all the specimens (indicated by a change in colour) was measured in the University chemical laboratory using an automatic recording spectrophotometer. The measurer did not know which specimens had been under the healer's influence as there was a 'double blind'. The results were carefully and statistically assessed and no paranormal effect was observed. So why had the American investigator achieved success?

We examined in detail the paper describing this and discovered that, amongst other procedures, there was a statistical flaw. The rate of haemolysis occurring in the experiment depended on the age of the blood samples, that is, how long they had been kept before use in an experiment. The American experiment had not taken into account this particular characteristic of blood and the organization of the experiment led to an apparently positive result. Books about blood and its properties do not appear to give any information about what happens to the rate of haemolysis with the length of time that a sample of blood has been kept. Haemotologists presumably take whatever tests they need on blood samples fairly soon and have no interest in variations with time of the kind we met. So the result of that experiment was merely a new and probably rather useless bit of extra information about blood!

THE BASIS OF EXPERIMENTS

Having considered a few examples of experiments which are carried out in parapsychological laboratories and elsewhere it is now time to say a little about the basis of experiment in science. The basis of science is human experience. When a number of human experiences appear to be falling into a pattern then a hypothesis — a model — can be devised which 'explains' those experiences. The purpose of the model is to achieve the capability of predicting future experiences in that same group. Experiments are devised to confirm or refute the model which has been constructed mentally. If the experiments confirm the validity of the model, the hypothesis, then it gradually assumes the status of a theory. If the results of experiments are to refute the model then those results provide additional facts to add to the others in order to construct a better model. It is important to understand that all the facts are facts of experience, that is, occurrences in the minds of human beings, and those same minds observe the patterns.

In testing a hypothesis certain factors are important. Clearly the first and most important factor is that irrelevancies should be excluded, that is, there should be no chance of any other possibility explaining a result — the experiment must eliminate this. Experiments should ideally be very simple and the simpler they are the better. The conditions under which the experiment is carried out must be carefully controlled so that variations in the conditions cannot lead to unanticipated results. And in parapsychological work the witnessing by reputable people is important because the sceptical realist or materialist will suggest that the experimenters are attempting to delude the scientific community or are indeed themselves deluded. An unusual result — and all the results of parapsychology might be described as unusual — must be verified many times.

It has been mentioned already (page 135) that in parapsychology all the relevant factors cannot be taken into account. In particular the mental state of all the people concerned is impossible to control. This is a major difficulty in parapsychology and one which occurs in other subjects only to a slight degree. Regarding the exclusion of irrelevancies and the control of the conditions of an experiment, variations of temperature and electrical voltage are evident parameters which must sometimes be closely controlled. If one is doing experiments on paranormal metal bending, where the bending is only exceedingly small, then delicate electrical measurements must be taken to detect it. This will often mean measuring very small voltages accurately. Voltages can be produced, of course, by unnoticed electrostatic effects: voltages can always be produced by unnoticed electrical interference and that necessitates such experiments being carried out in a screened room, designed to exclude such interference (radio waves, for example); it may be necessary to use a battery supply rather than the mains so that the supply system does not introduce unexpected variables. It is surely very clear that scientific laboratory work requires a particular training and experience and very often experiments require experts from a number of disciplines to be available for advice and comment.

KIRLIAN PHOTOGRAPHY

An amateur who is trying to carry out parapsychological experiments using apparatus with which he or she may not be completely familiar is likely to have to deal with many difficulties. A good example of this is provided by Kirlian photography. Kirlian photography is supposed by many people to produce a coloured picture of the aura or projecting portion of the etheric body, some people perhaps referring to the same thing as the astral body. (This is supposed to interpenetrate and project all round the physical body, and to indicate by its colours the state of health, as well as the emotional state.) Some Kirlian photographers give the most elaborate diagnoses of the physical/emotional and mental states of a subject from an inspection of the Kirlian photograph of a finger pad.

Kirlian photographs are obtained in the following way. For a finger-pad photograph, a flat metal electrode has placed on it a colour film on to which, in turn, is pressed a finger pad or a complete hand. A high-voltage, high-frequency discharge is allowed to take place between the electrode and the finger or hand, the latter being at earth potential. The voltage is high enough to produce an electrical discharge between the finger and the plate but the frequency is sufficiently high to eliminate discomfort felt by the subject. The photograph — which is not really a photograph at all — is produced by the electrical discharge which results from the high electric stresses in the air, producing ionization. The colour of the discharge depends on gases and other chemicals present and the electric stresses leading to the discharge depend on the curvature of the surfaces. The current flowing through the three emulsion layers of the film also decides features of the picture.

In another version of the process the discharge is caused to take place between the high-voltage plate electrode and an earthed plate electrode, a colour film being placed between them, together with an object on the emulsion side of the film.

Practitioners in this subject often suggest that the colours and brightness of the photograph indicate such things as vibrant health, a dull photograph indicating real or potential illness or disease. Dull photographs are apparently sometimes changed to bright ones after a 'healer' has given the ailing subject healing treatment. In the second arrangement described, if a leaf is used as the object then the practitioners suggest that the discharge from a living leaf is much brighter than the discharge from a dead leaf, for the same reasons that led to a bright finger-pad photograph — that there are ample quantities of 'vital subtle forces' flowing.

So how are the naïve practitioners misleading themselves and others? Is there anything at all in these claims?

What decides the details of the picture? These are determined by the flow of current through the three emulsion layers of the film together with the chemicals and gases (including moisture) in the air surrounding the Kirlian sandwich. Other features depend on the pressures between the components of the sandwich. In the case of a finger, the finger pressure determines the air spaces in which the discharges occur. Most importantly, the nature of the voltage, its variations with time, and the frequency, stability and the time of application of the voltage will all be of crucial importance. Deductions made from Kirlian photographs are based on comparisons between photographs. Thus a dull photograph before 'healing' followed by a bright one afterwards, may be taken as evidence for the efficacy of the healing. It should be quite clear that such comparisons are invalid and meaningless unless all the other variables are carefully controlled — and usually they are not. For example, finger pressure is not controlled and the high-voltage supply source is often a simple Tesla circuit which electrical engineers well know to be erratic. There are perhaps thirty different variables, the more important of which must be controlled for valid comparisons of photographs.

There is a well-known picture of a leaf (taken with the second Kirlian arrangement described) showing a discharge and a second picture taken with a section of the leaf cut out. Features in the space where the leaf has been removed are suggested as indicating the presence still of the 'etheric material'. In the case of the leaf picture, it seems highly likely that the discharge in the space where the piece was cut out was probably due to moisture and chemical substances left on the plate as a result of not cleaning the electrodes properly. (A competent experimenter would have given details in his or her paper of how the electrodes were cleaned. Such details were significant by their absence.) The difference in the brightness of discharge between that from a living leaf and that from a dead leaf is merely due to the moisture present in the living leaf.

There is no real mystery about any of this to an electrical engineer who is familiar with high-voltage discharges. To someone who is not familiar with electrical engineering, or indeed with any physics at all, the electrical discharge is just as mysterious as is the etheric body. Because the way in which the etheric body appears to psychic 'vision' round the surface of a physical body is rather like the way an electrical discharge appears around the surface of a discharging object, then a comparison is often drawn between them.

I well remember a lecture on Kirlian photography in which it was described how two people who barely knew each other had placed their forefingers

By using electrodes and a high-voltage high-frequency discharge, Kirlian photographs are produced on colour film. Some people believe this gives an impression of the etheric form and that the discharge from a living healthy leaf is brighter than that from a dead leaf. However, it is difficult to control all the variables and make valid comparisons.

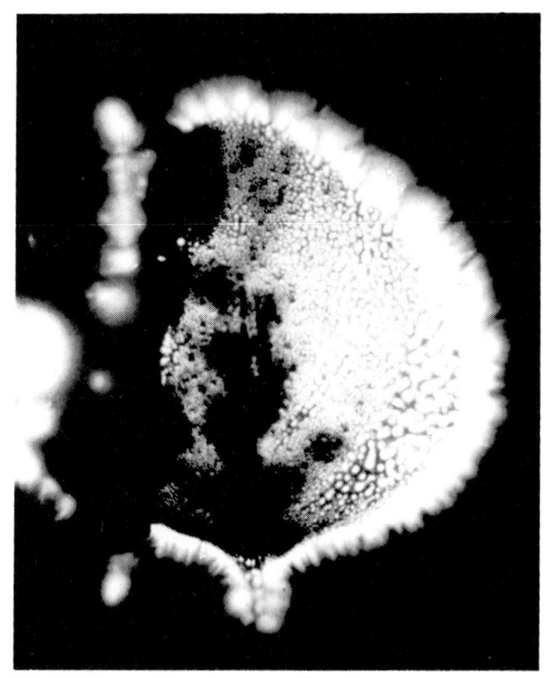

The discharge from a normal leaf.

Kirlian apparatus employs this 'sandwich' arrangement. When an object such as a leaf is to be photographed, it is placed between the upper (earthed) plate and the film surface.

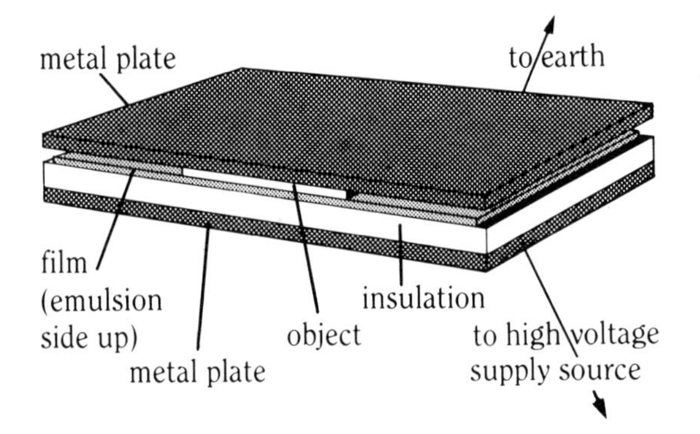

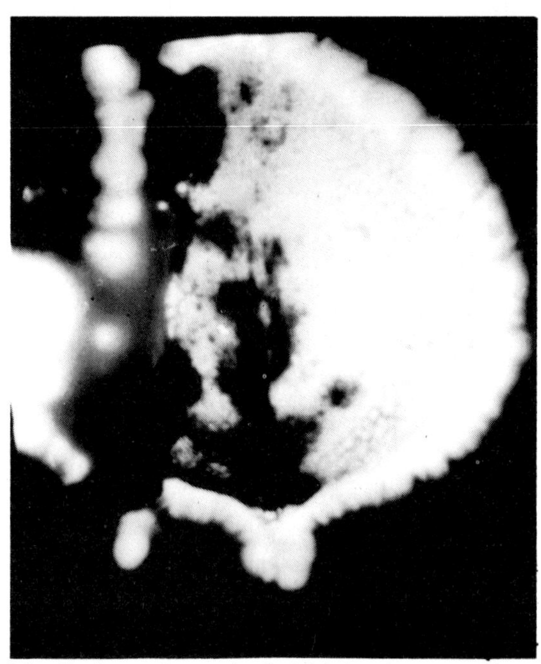

The discharge from a leaf which has been cut. Do features in the space where a piece of leaf has been removed indicate etheric material?

When a fingertip, hand or some other part of the body is photographed, no upper electrode is required since the subject is, in normal conditions, earthed. If the participant is not properly earthed, however – perhaps because he or she is wearing shoes with rubber soles, or is on a carpet with a rubber underlay – there may be a loss of quality in the Kirlian picture obtained.

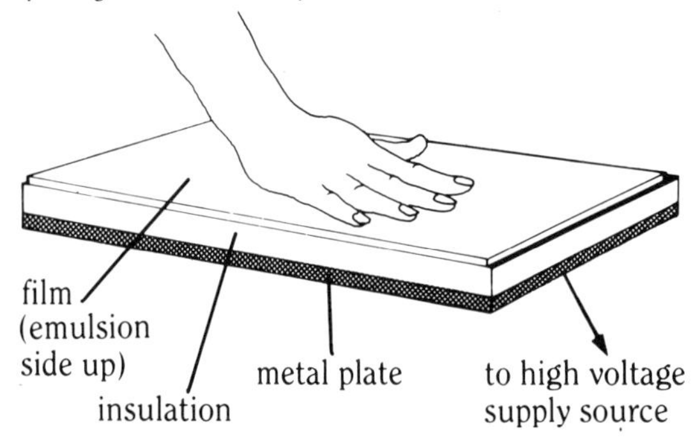

near to each other on the Kirlian sandwich, a photograph being taken. Then two people who were engaged to be married placed their fingers close together in a similar way. The lecturer pointed out that the etheric material was merging together in the case of the engaged couple! When I asked him whether there were any other photographs of the discharges taken, both for the two people who did not know each other and for the engaged couple, he said that there were but he had selected the pair which showed the effect best. Again, it will readily be appreciated that the audience should have been shown all the photographs and not only the two which showed what the lecturer was suggesting. As he had no control over any of the numerous variables no valid deductions of any kind could be drawn from the two photographs shown.

Kirlian photography has also been used for medical diagnosis. L.W. Konikiewicz, an American researcher, used a stable oscillator to produce the high-voltage supply and controlled all the independent variables. He used an environmental chamber to maintain the humidity at a constant value and he took special care to keep the film flat. He cleansed the fingers of his subjects with distilled water, drying them in air and covering them with cotton gloves up to the moment of the test to prevent contamination. He controlled the finger pressure carefully, and the time of the discharge. Using double-blind studies he correctly identified with a high order of accuracy cystic fibrosis patients and carriers of the gene.

Changes in Kirlian photographs (assuming all the independent variables to be properly controlled) before and after healing should be expected as the chemical constituents of the sweat on the skin will be altered. It is well known that calmness and relaxation alone alter these. This has nothing to do with subtle bodies and subtle forces unknown to science. Kirlian photography is potentially useful if all the important variables are properly controlled and the work is undertaken by professionally trained workers. Otherwise it provides yet another way by which the ill-informed may deceive both themselves and their subjects.

One final remark – as psychokinesis sometimes genuinely takes place then it might be possible for the electrical discharge of a Kirlian photograph to be operated upon and modified by the minds of people such as healers, their Georges indicating to them appropriate diseases present by means of such changes. This suggestion is similar to the explanation of the dark patches seen by Kilner using his Kilner screens (see page 100). However, such events are not likely to occur very frequently!

Something which is very important – but which is almost always omitted from descriptions of scientific experiments in parapsychology – is the mind of the subject. Perhaps of equal importance, but similarly overlooked, are the minds of the experimenters. If a subject achieves remarkably good results in a telepathy experiment when using ganzfeld (see page 146) or equally good results when trying to influence Jahn's electrical-pulse equipment, then it is surely of great importance that the results should clearly describe exactly what those successful subjects were doing with their minds. This vital piece of information is almost always omitted from the reports.

THE PHILIP EXPERIMENT (COMMUNICATION WITH AN IMAGINARY GHOST)

The Philip experiment has been referred to several times earlier and provides an example of an outstandingly successful experiment. Its description follows naturally from the previous paragraph, for what the experimenters were doing with their minds was of great importance. The Philip experiment was carried out in Toronto by a group of researchers headed by Dr George Owen (who was once a Council Member of The Society for Psychical Research). Other members of the group included Mrs Iris Owen and various members of their psychical research society including a psychiatrist. The purpose of their experiment was to see whether they could establish communication with someone they had brought into existence by imagination only.

The group wrote a fictitious story set in England

several hundred years ago and involving a stately home, the history of which is well known. The story they devised involved a nobleman named Philip, his frigid wife and a beautiful gypsy with whom Philip was enamoured. Philip installed his gypsy lover in a cottage on his estate. His wife discovered her and accused her of being a witch. She was consequently burned at the stake. Philip, who had done nothing

Portrait of Philip, the subject of a fascinating experiment.

to save her, was tortured by remorse and eventually killed himself. The Owen group read history books concerned with the period in which they had set their story, and they even made a sketch of their imaginary character Philip. They also obtained photographs of the site.

When all this was clear in their minds they sat round a table and tried to communicate with Philip, who had of course, in their story, died several hundred years ago. They had very little success until they tried changing the solemn 'scientific' atmosphere and kept their séances light and cheerful, making jokes and singing. When they changed the atmosphere in this way they very quickly obtained contact with Philip who produced raps on the table on which they had all placed their fingertips. Philip used a code on which they had agreed, giving one rap for YES and two for NO. Philip, who was imagined to be standing near the table in his astral body, produced very positive raps. He answered all sorts of questions which confirmed the facts they had decided upon in their story and occasionally he extended it, the extensions not always being historically accurate. Philip's opinions tended to depend on whom was present. For example, he had no objection to smoking when the only smoker in the group was present; at other times he disliked it.

Philip's raps were quite clearly paranormal, that is, no one was consciously or unconsciously deliberately rapping the table. This was perfectly clear: the light was always good and the table top was a thin round disc which was mounted on a narrow central post. When a vibration transducer was attached to the surface of the table the displacement of the table varied with time in a way which was the reverse of what would have been obtained if someone had physically rapped the table. (If a table is rapped the surface is displaced at once to the maximum and then restores itself to its original equilibrium position over a short time. The paranormal raps produced a more gradual build-up to the maximum displacement.)

A most interesting feature of this experiment occurred during one séance where the raps were fairly weak. The group said, 'Come on, Philip, you can do better than this. If you don't produce stronger raps we can always send you away. We only made you up, you know.' The raps then practically died away and the group had to strengthen Philip's identity by clear imagination in order to build the raps back up to a reasonable level again.

On one occasion, after a successful visit by Matthew Manning, who did, amongst other things, some 'spoon bending', the group asked Philip if he could bend a key left on the table in the locked séance room during the night. George Owen described how, when he entered in the morning, the key was indeed bent.

This production of physical energy — and clearly the physical raps involved the deployment of physical energy which was certainly not being consciously expended by anyone present — is of fundamental importance to science. It has been suggested — and I certainly agree — that if a Nobel prize were awarded for a parapsychological experiment then this should win it.

The phenomena I have so briefly described were always obtained in a good light and were strong enough to occur in the presence of more than one sceptical observer. They have been photographed and have appeared in radio and television programmes a number of times. The group was kind enough to allow me to sit in on a séance when I visited Toronto and I have not the slightest doubt that everything was genuine and took place just as I have described.

Originally the group hoped to be able actually to see Philip. They had no success with this but who knows what might have occurred had they continued with the experiment! (After some years they all became somewhat bored by it.) The possibility of seeing the imaginary Philip as well as 'hearing' him and so communicating with him does not seem to be at all unlikely. Hallucinatory experiences are by no means uncommon. An intriguing thought occurred to me when I read the original papers on the Philip experiment. There are, I believe, lots of groups practising black magic in various of our communities and probably trying to 'raise the devil'. If those groups had the clarity of thought and the persistence of the Toronto group it would not surprise me in the least if they were not in the same way able to communicate with the devil. However, the devil would of course have been created by themselves: it would, like Philip, be a thought form but none the less able to deploy physical energy. Maybe they would even see him, complete with horns and a forked tail! Who knows what dark and undesirable events might follow after raising from the unconscious an archetype of that highly unpleasant kind.

THE VIEWS OF BATCHELDOR

The Owen group had read the views of an English researcher, Batcheldor, before they tried their experiments so a few words outlining his ideas are appropriate here. Batcheldor holds the view that psychokinesis derives from a certain state of belief or expectancy and that this belief in the likely occurrence of PK might be created by the agent's misconstruing some normal event as being due to PK. Batcheldor's methods involve exposing hopeful PK agents to such events — which they think are examples of PK — and then it is hoped that genuine PK will follow. In other words Batcheldor suggests that there is in existence an emotional resistance to PK which has to be evaded in order to produce PK occurrences.

Batcheldor and his collaborators used tables around which a group would sit in a darkened room. If a chosen member of the group, unknown to the others, raised the table in the ordinary way and not by PK then instrumentation attached to the table would indicate this. If the result of the table rising (or raps taking place) was to cause a change in the belief of members of the circle present then it was hoped that afterwards the table would rise (or raps would be produced) paranormally. Meanwhile the instrumentation would indicate quite clearly what had been occurring.

I have earlier suggested that perhaps children, more often than adults, can produce paranormal spoon bending *because they do not know that it is impossible,* that is, that our current scientific models forbid it. The aim of the Batcheldor scheme is, it will now be clear, to change (by a device) the belief structure of the participants. If a new paradigm for the operation of the physical world, involving different laws and possibilities, can be produced, even if only temporarily, then the theory is that certain new phenomena become possible. The results of the Philip experiment show that, at least in the conditions of that experiment, Batcheldor is right. One is reminded of the Christian injunction that a strong enough belief is capable of 'moving mountains'. This is perhaps more than a metaphor — at least the evidence appears to indicate that it is worthy of some deeper consideration than is normally given to it by scientific people.

A PARAPSYCHOLOGICAL LABORATORY

From what has been said so far, the sort of features which a good parapsychology laboratory should have will be evident. Clearly, it should have a warm, friendly and welcoming atmosphere with carpets on the floor and pleasing furniture. Ideally it should have the appearance of a comfortable home. In addition it will be helpful if it is peaceful and quiet with subdued lighting and flowers, and with a nearby kitchen for providing refreshments. The whole aim should be to make the participants in a test feel welcome and at ease.

Regarding the scientific equipment which should be available, means for measuring and recording electrical brain rhythms and other physiological parameters should be to hand. It may be difficult to arrange for such equipment to blend in with the furniture but the laboratory designers must do their best, perhaps using screens or curtains so that the equipment does not dominate the room. The general atmosphere of the laboratory should be primarily for the purpose of allowing the subjects to go into whatever altered states of consciousness seem appropriate to them and the experimenters. (Developing psi by hypnosis has sometimes proved effective.) As all kinds of measurements will be needed — from the physiological ones already mentioned to a whole range concerned with engineering and other aspects of applied physics, chemistry and so on — more normal university laboratories should not be too far away so that measurements can either be made there, perhaps using remote control, or equipment can be brought in as necessary. It is clear that such a laboratory should be within (or at least, not too far away from) a broadly based university which encompasses all the relevant disciplines. It is important also for experiments in psi to have another similar pleasant laboratory within which a second subject and experimenter can be installed for experiments between two subjects, with appropriate communication arrangements between the two rooms.

GANZFELD EXPERIMENTS

In our model, because George is endeavouring in ESP experiments to send subtle information floating up into the conscious mind, it would appear to be helpful if the conscious mind is not already occupied with normal thoughts. The apparent desirability of a dreamy dissociated state has been mentioned earlier. This is achieved in the ganzfeld (which is German for whole field) by arranging for the subject to lie comfortably with half table-tennis balls taped over the eyes, sealed in place with cotton wool, and a red light shining on them to create a uniform visual field. Ear phones provide the subject with 'white noise' — rather like the rushing sound of the sea — conveying no information. In this way all information input from the five senses is reduced to a low level and subjects normally begin quite soon to experience imagery. This is sometimes extrasensory in origin; for example, it may emanate from an agent in another room who is looking at a random selection of pictures. The subject gives continuous reports of all thoughts, images and feelings during the ganzfeld stimulation.

In an early experiment (1975) an agent gazed at a picture in another room while a ganzfeld group and a control group not in the ganzfeld gave their impressions. The ganzfeld group scored significantly better than the control group.

PERSUADING PSYCHIC SUBJECTS TO BE TESTED

No matter how welcoming the laboratory and friendly the researcher, it is sometimes difficult to persuade an ostensibly good psychic to agree to take part in experiments. There seem to be two reasons for this. One reason is that psychics are only too well aware that the faculties are not under conscious control and cannot always be produced at will. They are nervous that test conditions will perhaps inhibit them and that the researcher will assume that they can never exhibit the powers they claim and cease the experiments too soon. Their reputation will then be gone and their livelihood with it. Such likely subjects must be reassured by explaining that the researcher is well aware of the possible difficulties and will not discontinue the experiments too soon. Also, the experiments will be carefully planned, in collaboration with the subject so far as is scientifically permissible (that is, without spoiling their scientific validity) so that the subject is happy with them. And if the results are negative then discretion will be used.

The second reason for hesitancy, shown by such well-known subjects as Uri Geller and Matthew Manning, stems from their experience that investigators whom they understand to have achieved good results do not always promptly, or ever at all, publish them. This is more difficult. In such a complex and unusual subject as parapsychology a carefully thought-out protocol for an experiment might prove, afterwards, to have contained hidden snags which would either invalidate it scientifically or indicate the inexperience of the investigator; there might be 'normal' explanations for apparently successful results. A well-known psychic does not want his or her time wasted by inadequate researchers (who may have reputations in other subjects but perhaps are not particularly competent in parapsychology). Successful psychics may find more profitable outlets for their talents. It is reported that Uri Geller is a multi-millionaire as a result of using his psychic talents to serve industry — for instance by finding minerals by divining.

The moral for a researcher is, of course, to realize fully the potential difficulties and organize a 'steering committee' for the experiments, incorporating all the necessary experience and all the essential disciplines. It is difficult to see what else can be done, but a happy subject is probably an essential prerequisite to success.

HYPOTHESES THAT CANNOT BE TESTED

There is one final matter that must be mentioned in connection with scientific experiments with psychics. Psychics are not necessarily scientific people and so they may not understand the scientific method. They, and their non-scientific collaborators, frequently produce information from 'scientific communicators' which looks to them, the receivers of the communications, like scientific material but is in fact nothing of the kind. I have many times been sent long statements from 'Sir Oliver Lodge' or 'Sir William Crookes' which not only cannot be tested for validity but which indeed have very little clear meaning at all. A statement which cannot in principle be refuted (or confirmed) by tests — or reasoning from earlier tests — is not a scientific statement and of very little interest to a scientist because nothing can be done with it. (If you believe it and follow it you are being 'religious', not 'scientific'.)

Many are the letters I have had from the leader of a 'circle' telling me that a great scientist informed the circle that he had 'come on a very high vibration' to inform the world of science that on some high spiritual plane certain scientific facts were being transmitted to the earth by a group of scientists. I have never yet received any such communication which had a meaning I could follow or test. For a start, what exactly is supposed to be vibrating?

It looks very much as though earthly scientists can receive help in only the most general sense from paranormal sources! This may be (how can we know?) because a medium's George does not have either the language or the concepts to transmit detailed information. Scientists will have to develop psychic faculties in themselves! I have already described several cases of scientists having hypnagogic imagery of great help to them. They had already made the preparations in their own minds, and reaped the result. But perhaps it was another higher part of themselves which was so helpful.

Following intuition or a 'hunch' may not be pure coincidence. New developments depend upon new ideas and concepts. Perhaps the seeds of such initiatives are growing deep down in the unconscious mind; they might be more quickly brought to the conscious mind by implementing the discoveries of parapsychology.

12
WHAT DO WE REALLY KNOW?

SHIFTING PARADIGMS

So where have we reached? People often say to me, 'In view of all these experiences you have had and all the thought you must have given to them, what do you now think? What do you consider to be reasonable deductions to draw from all this? Is there anything that a rational human being, with reasonable confidence that he or she is on the right lines, should do?' Such questioners are often well aware that nothing in this universe is 'provable'. We cannot be one hundred per cent sure of anything at all. Nonetheless, we must proceed with our lives on the basis of a reasonable degree of probability that we are doing the right things. So it is possible to be just a little speculative, based on reasonable probabilities.

I believe that there is in fact a 'paradigm shift' on the way. Current Western culture has as its basis of belief and perception the paradigm that the universe is all 'out there', distributed in three-dimensional space, with time steadily flowing, and we can observe it through our five senses, and in no other way, without affecting it. And what we observe through our senses, provided we are alert to possible misinterpretation, is just what it is.

This is the paradigm which I think will, before very long, be rejected in favour of another. (The change of paradigm has, in fact, already started in particle physics.)

Of course every culture has had its paradigms and every culture, I imagine, has felt that theirs is closer to the truth than any other. It was the great scientist Lord Kelvin who said that in his view most of the scientific problems were, by and large, solved and that scientists just had a little tidying up to do. He suggested that if we knew the mass and velocity of every particle in the universe then we should know in principle the whole future of the universe. Then along came Planck, Einstein, Schrödinger, Heisenberg, Bohr and the others! No scientist would dare say such a thing today: there has been a shift of paradigm.

I very much hope that it is clear by now that, in my view, scientists are not describing the physical world out there with increasing accuracy but, on the contrary, science is the process of building mental models to represent our experiences. Those experiences are in our minds (whatever that may mean — but it is all we have); and the body, with its five senses, is part of this world that we have 'postulated' to make sense of our experiences. Lord Kelvin's paradigm described in the previous paragraph is an example of such a mental model — one which changes, and is clearly not an objective description of the physical world. I believe that there is no such thing as an objective description of anything: we have these experiences only in our minds, and nothing else. Those who stoutly maintain that there are all sorts of things all around us and that they are independent of us are indicating, to me at least, that they are naïve realists. Of course we are all well aware that we have to behave for most of the time as though we are naïve realists or life would become impossible. But the basis of this book is an attempt to have a deeper understanding of our experiences, which will encompass and not ignore as though they did not exist, all the anomalous experiences which I have been describing in the preceding chapters.

Before leaving the question of mental models representing experience it might be a good idea to consider one more not previously mentioned. The earlier example was of Newton's idea of masses attracting each other — upon which his law of gravitation was based. Because more accurate measurement showed this not to be quite correct, Einstein put forward his quite different model involving masses distorting the space-time continuum — with no idea of their attracting each other. Another similar example is that of phlogiston which, some two-hundred years ago, chemists considered to be released during combustion from burning materials. Evidence gradually accumulated of phenomena which could not be explained with the phlogiston theory except with great difficulty. The 'discoverer' of oxygen, Priestley, did not accept the fact that he had done this: he believed that he had isolated air with the phlogiston removed – which he called 'dephlogisticated air'. His paradigm completely prevented him from observing the substance we now call oxygen. However, once a paradigm shift is over the world is changed for those scientists who accept the new structure. Thus, earlier chemists considered that combustion released phlogiston; later chemists considered that combustion consumed oxygen.

The same transition of ideas occurred with our understanding of the sun's position. Aristotle considered the sun to revolve around the earth; sensible scientists like Ptolemy considered this perfectly obvious and 'proved' it. Copernicus suggested that, on the contrary, the sun was the centre of the system and Galileo learned how to build telescopes which enabled him to 'verify' this model. Many people refused to look through Galileo's telescopes – because it was 'perfectly obvious' that the sun revolved around the earth. Such non-questions as whether or not the earth revolved around the sun were unfit for the concern of a respectable scientist.

The situation was exactly as it is today in regard to the evidence for parapsychological phenomena described earlier in this book. After Copernicus and Galileo the earth revolved around the sun. The change involved instability and disorder as the old concepts broke down and were replaced by the new.

Some older scientists were unable to make the switch. You will remember that optical illusion in which you see two faces in profile and suddenly they change to a vase. There is another one in which you are looking at the tops of stairs going upwards when suddenly there is a change and you find you are looking at the stairs from underneath. Nothing has changed except your perception, that is, your mind.

If the model through which the perception forms does not contain appropriate categories to allow recognition of an event, that event may be unrecognized or misinterpreted, even in the face of considerable evidence. We can fail to see what we do not expect and we may see something which is not there at all. 'Perception' is filtered and organized through our models. Science is a form of perception and gives sense and organization to what we see. It provides no unique access to truth.

TODAY'S SCIENTISTS

After changes of paradigm, rewriting begins and prospective scientists are taught along the new lines. The public derives its knowledge of science from the scientists and the new perspective is considered then to be the correct one. It is interesting that no one now considers Newton's interests in alchemy: a 'good' scientist considers them to be merely medieval superstitions. The younger generations of scientists think that science is both orderly and progressive; they are taught that the older scientists made imperfect attempts to represent the scene from the current viewpoint rather than appreciating that these scientists had a different perspective which has now been rejected. In other words, scientific training is not broadening but, quite to the contrary, it is limiting. Young scientists learn to see things in one particular way and no other and consider that they know what the truth is, rejecting older ideas about the world because they conflict with this knowledge.

The point just made is of enormous importance because present paradigms are used by scientists to determine what qualifies as fact and how it should be observed and measured. In other words, the paradigm provides order and enables predictions to be made – but only from a particular point of view. It is absolutely wrong for scientists to consider, as most do, that they are detached objective observers describing what goes on. Most, certainly, do not appreciate (unless, perhaps, they are good particle physicists) that they are themselves a part of the process. Everything, they consider, would occur in just the same way if they were not there. They are concerned solely with things which are repeatable, that is, that others following the same procedure should obtain the same results. Such data are said to be facts.

The approach of modern scientists is reductionist: everything is looked at all by itself and complex phenomena seen as nothing more than the total of simple interactions.

So what are the factors which are ignored and considered as non-important in this modern reductionist picture of the universe? They are such things as beliefs, attitudes or intentions, which serious scientists, considering themselves to be detached, critical objective observers, do not consider worthy of notice because they are not within their particular paradigm. Phenomena depending on these factors are then apparently unrepeatable, unstable and erratic. However, they cannot be made not to exist! The very word paranormal means that the phenomena do not fit our current paradigms; that they do not conform to present definitions of the normal and what is allowed to exist or make sense.

The greatest, most noble members of humanity have had both psychic and mystical experiences. People who have had mystical experiences consider them to be much more real than their ordinary experiences of the physical world. None the less, the modern scientist would say that this was all delusion and would take absolutely no notice of it — as such experiences do not fit the current paradigm.

THE EFFECTS OF LYSERGIC ACID (LSD)

Well over twenty years ago some scientific workers studied the mystical experiences of one hundred and forty-seven subjects obtained by use of the hallucinogenic drug lysergic acid. (It was necessary, they explained, to give the drug several times in order to 'stabilize' the experience, otherwise they found that the subjects were completely disorientated and did not produce consistent statements.) In a large majority of cases, after the experience had been stabilized the subjects discovered (and other researchers have found the same) a feeling of unity with the rest of the universe: in other words, they stated that there was only one life and that it was the source of the good, the true and the beautiful. They had to make changes to their concepts of self and body. If they felt that there was only one life they could no longer view themselves as individual bodies. They had changes in their perceptions of space and time and some of them referred to the fact that they could be anywhere in the universe they wished at any time — past, present or future. They spoke of being in the 'eternal now'. Many of the subjects had a feeling of a profound understanding of the sort of problems with which philosophy and religion deal. They all had an empathy with the others and a much wider range of emotions.

Such experiences would be, to many psychologists, hallucinations, and technically of course they were because the perceptions were quite different from the normal perceptions of the so-called physical world. However, to the subjects they appeared much more real than normal life. In fact some of them suggested that coming back to normal consciousness after the joy, bliss, light and unity of that level of consciousness was rather like diving down a deep, dark gloomy well and becoming embedded in the mud at the bottom. Poetry and religion, and all of the greatest art forms, appear to be based on the mystical experience.

For all that, these greatest and most enobling experiences of humanity cut absolutely no ice with many modern scientists, particularly behaviouristic psychologists, for whom they are just hallucinations — the product of a mind disordered by a drug or by religious and meditational practices. However, surely they are part of the general data which we must consider when trying to understand ourselves and the nature of the universe. In the iceberg diagram, mystical experience — an experience of a transcendent unity without the concepts of space and time — would be beyond the bottom of the iceberg, where it is merging into the sea.

WHAT IS IT LIKE AFTER DEATH?

Earlier, I indicated that a little more about human survival of bodily death would be found in the final chapter. Professor H.H. Price produced a thoughtful and stimulating paper some years ago in which he deduced, from philosophical and psychological considerations, the sort of life one might expect after the death of the body, assuming we still existed. His deductions agree very well not only with the ostensible information through mediums and the near-death experience but also with the teachings of Hinduism.

Clearly after death we would have no information coming in via the senses and therefore no sense perceptions. How could we experience a world? It would surely be a kind of dream world: when asleep we have no sensory input but still have experiences. After death George would really come into his own and produce objects of awareness about which we could have thoughts, desires and emotions. The next world would, on this argument, be a world of mental images. It would be quite solid – there is nothing imaginary about mental images. Sometimes objects would behave in a queer way but this would not be too disconcerting and our identity would not be broken. An image world would be, to those who experienced it, just as real as this one; in fact they might have difficulty in realizing they were dead. It would be a perfectly good world, in which one would feel completely alive. But what about a body? (Saint Paul had that difficulty! 'How are the dead raised up and in what body do they come?') They could have images representing the body they had here. They might find that their image bodies also were subject to peculiar causal laws in that wishes tended automatically to fulfil themselves. The body could be young and vigorous and dressed in any way that its owner wished. A wish to go to New York City might be followed at once by a set of New-York-City-style images and the owner of the body would realize that 'going somewhere' was a little different. He or she might conclude that the body was not the same as the physical body and might call it a psychic or spiritual body, very like the old one but having different properties.

We have earlier explained, and shown by means of an experiment (see page 97), that this body (and this world) would not be in physical space but in its own space. Passing from the physical world to the next might be thought of as a change of consciousness, like waking to a dream. It would involve a change from a perceptual consciousness to an imaging type of consciousness. You might suggest that discarnate minds are thus in a state of perpetual delusion. If so then they would have to put up with it. However, we say delusion only because the experiences would be different from those of the physical world. People in the next world would be deluded only in the sense that their bodies and their world were not really physical though they might mistakenly think so. But it is another world, as it should be, having different space and laws.

Would a world of mental images be private? Not if we accept telepathy. Telepathy might be more common in a disembodied state than it is here and the image world might be the joint product of a group of telepathically interacting minds and public to them all. It would not have unrestricted public access as it seems likely there would be many such worlds for each group of like-minded personalities. There are a number of groups of people having rather narrow religious ideas possessing very clear views as to what 'heaven' is like. They would discover that they were in such a world, with all the 'unbelievers' being excluded. ('In My Father's house are many mansions.')

Such a world would be mind-dependent. It would be dependent on the memories and desires of those who experienced it; memory, as Price puts it, providing the pigments and desires painting the picture. Desires unsatisfied in earthly life might play an important part. This could seem agreeable, but desires repressed because they were too painful or disagreeable to admit might also be important, and the same might be true of repressed memories.

Such an after-death world of mental images has been well described by Hindu thinkers as *kama loka* or 'world of desire'.

Material possessions clearly cannot be taken into the next world – but this would be no loss because if they were remembered well enough, image replicas could be produced of them.

Price mentions the scathing remarks of some

people concerning the materialistic character of what comes through mediums. Agreeable houses, beautiful landscapes and gardens are described. They argue apparently that this materialistic character is evidence against its genuineness. On the contrary, this is evidence *for* its genuineness as most people do like material objects and are deeply interested in them. If the objectors are saying that such a world is not worth having and would prefer a different state because they find this uninspired and unsatisfying — then they will indeed experience something different. A mind-dependent world would tend to be a wish-fulfilment world.

In case you think this is too good to be true it is easy to argue that it is the reverse. There would be many next worlds, not just one, and the world each of us could expect after death would depend on the kind of person we are. It is easy to see that some people's next world would be more like purgatory than paradise because they have conflicting desires. Few of us are completely integrated personalities and, Price suggests, the one-pointed saint probably comes the closest. Sometimes when our desires appear to be relatively harmonious, appearances can be deceptive. Conscious desires are not in apparent conflict but this may well have been achieved at the cost of repression. There are unconscious desires conflicting with the neatly organized pattern of conscious life: the seeming harmony might vanish after a person is dead. Formally repressed desires would be manifested by appropriate images, and they might be horrifying — as some dream images are for the same reasons. (The 'secrets of the heart' would be revealed.) They would certainly be wish-fulfilment images but the wishes would be in conflict with others and the resulting emotional state might be worse than the worse nightmare — worse in that the subjects are unable to wake up. They might find themselves doing cruel actions which they never did in earthly life: but the desires would have been there even though repressed and unacknowledged. Cruel desires would fulfil themselves by creating appropriate images. But unfortunately for their comfort their personality encompasses benevolent desires too so they are distressed and horrified even though, in a sense, the situations are expressions of their own desires. For instance, psychoanalysts tell us that there is often an underlying urge to be punished — the result of guilty repressed feelings.

It is clear that such unpleasant experiences would not be literally punishments. They would be inflicted by no external judge but each person's purgatory would be just the automatic consequences of his or her own desires. The life after death, on these arguments, would be an expression of what each person truly is — it will all depend on what we have made of ourselves during earthly life.

At first sight one might think that an image world contained no hard facts and so there was nothing objective. However, a man's or woman's character *is* objective in that it exists whether we like it or not.

The next world as pictured would be subject to law but not to the laws of physics: such laws might be more like the laws of psychology. If we dislike the image world our memories and desires create for us — if when we get what we want we are horrified — we have to set about altering our character, and this might be a long and painful process.

Some people say all desires, even permanent and habitual ones, would wear out in time by the mere process of being satisfied. In such cases this dream-like purgatory would be only temporary but it would be where we are brought face-to-face with what we really are. When that has perhaps become a thing of the past — what would the world after that be like? Would we still have personal identity when we are not even dreaming? Price puts the question: would the soul become something greater? What sort of experiences could we imagine for it? He refers us to the mystics, who have tried to describe (in allegory, because there is nothing equivalent here) this higher state of consciousness — which they experience while in the physical body.

Price's account of the next world is, you will appreciate, in some respects not at all unlike what we have in effect suggested this world may be like. We postulate a physical body having five senses but actually all we really know is that we have experiences, images, in the mind. Those images are all we have. 'Material objects' are just collections of ideas.

Price wrote his paper along these lines before the information about near-death experiences was available. So far as they go, these experiences appear to confirm his views. Scriptural writings appear to do the same. I do not think that Price was far wrong in his descriptions of the hypothetical 'next world'. Certainly they agree very well with the evidence presented earlier in this book. Perhaps, if we think there is something significant and possibly important and true in these views, it might be a good idea to start to improve our characters right now! It could be much more painful to let the next world do it for us!

THE VIEWS OF BOHM, PRIBRAM AND SHELDRAKE

It will be helpful at this stage to consider briefly the views of these three writers. In this way we may be better able to develop some kind of conception of the problems presented by the data we have considered — especially those which are not consistent with the current Western scientific model of the universe.

David Bohm

David Bohm is a professor of theoretical physics at the University of London, and was a friend of the philosophical religious teacher Krishnamurti. His ideas — some of which are very like those to be found in Buddhism and certain aspects of Hinduism — could well be the basis of the new paradigm for the West, integrating Western science in all its aspects, including the data of parapsychology, with religion. (By religion I refer to mystical experience rather than to the ritual practices of the orthodox religions of the world.)

A tentative understanding of what Bohm's theory is all about is best attained by considering the hologram. The hologram is produced in the following way. Coherent (laser) light (that is light in which all the wave fronts are in phase) is allowed to fall on to a half-silvered mirror. Some of the light passes through the mirror on to a photographic plate and the remaining (and reflected) light strikes an object, every part of the object producing wave fronts. The wave fronts also strike the photographic plate, interfering with the wave fronts arriving directly, and an 'interference pattern' is produced. This is to the eye merely masses of confused whirls. (The sort of interference pattern between the waves produced by separate stones dropped into a pond is similar.) These confused whirls have, however, an orderly pattern and this is called a hologram. Thus the photographic plate has on it a pattern, a blur, with order. If coherent light is now shone through the photographic plate it produces an image of the original solid object in space. This image can be looked at from different angles as though it too were solid. If a small piece of the photographic plate is cut off, then the whole object is still reproduced but with less detail and from fewer angles. In other words, a tiny bit of the hologram can still reproduce the whole object and there is no one-to-one correspondence between points on the object and points on the hologram. Bohm suggests that the universe is like a hologram, every part having in it every other part, as the mystics say.

Bohm has a very good analogy to his holographic universe. He imagines a vessel consisting of two concentric glass cylinders having a viscous

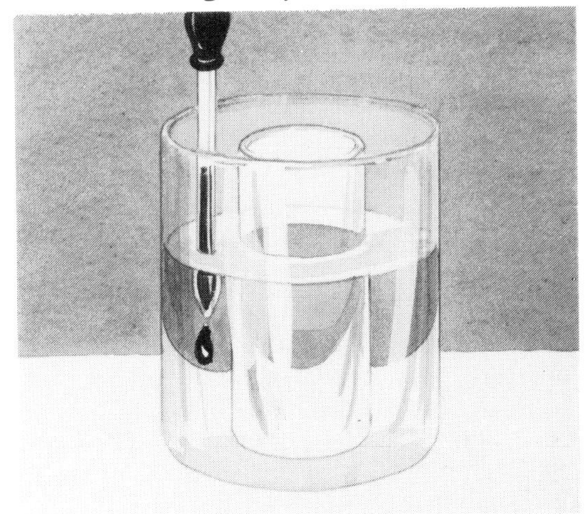

liquid between them (like glycerine, or treacle). The outer cylinder can be rotated slowly to stir the liquid gently so that there is no diffusion. Now imagine that an insoluble droplet of ink is put in the glycerine by an eye dropper and the outer cylinder stirred n times. The ink droplet will be drawn out into a fine thread distributed throughout the system, enfolded in it and made invisible — or seen only as a greyish coloration. If the stirring device is now reversed, after n revolutions the droplet will have coalesced again and will suddenly become visible, having been unfolded — disappearing again if the stirring is continued. If now, after the first drop is added and stirred n times, a second drop is added near where the first drop was added and the stirring is continued n more times — and then a third droplet placed near the second, stirring again being continued for n times — and so on, all of the droplets will disappear. Each will be all over the liquid — enfolded in it. Imagine now, Bohm says, that the stirring is reversed and done quickly. The first ink droplet will reappear and disappear and then the second will do the same near it, and then the third. Persistence of vision will make it look as though a real ink droplet is moving through space, in time, even though there is no such single object.

In just this way, says Bohm, all apparent substance and movement of the world are illusory. They emerge from another more primary order of the universe. Bohm calls this phenomenon the holomovement. It is imponderable and indefinable. Our ordinary science gives theories which are only approximations for local conditions. We look at nature through lenses and try to objectify and so alter what we hope to see. Nature will not stand still and display its edges because its true nature is in another order of reality where there are no things. Electrons sometimes look like particles, sometimes like waves. They jump from one point to another, apparently without crossing the space between. We are trying to bring the 'observed' into focus and running into difficulties. The blur is a more accurate representation: the blur is the basic reality. In Bohm's cylindrical analogy you could imagine what would happen to the 'travelling ink spot' if an object were placed in its way. It would not behave like a moving ink spot at all but 'waves' would be produced. It is just like that with an electron.

In particle physics, particles sometimes appear to jump from one point to another point without crossing the space in between. Bohm's analogy would show the same: if a magnified high-speed movie film were taken of the 'moving ink spot' it would appear to jump from one place to another without crossing the space in between.

That is a very brief and inadequate sketch of Bohm's idea that all apparent substance and movement are illusory — as it is called in Sanskrit, a *maya*. Everything is enfolded in everything else.

Pribram

Karl Pribram is a neurologist and a professor in the University of Stanford, California. Pribram had been trying for years to discover by experiments on brains how the brain stores memory and had found, as have others, that a particular memory cannot be removed by cutting out small pieces of the brain. It is as though the memory is spread all over the brain. Cutting away bits does not destroy the memory but weakens it and spoils detail. The brain itself seems to be like a three-dimensional hologram in which information is stored; Pribram suggests this is stored in the form of proteins (rather as silver oxide stores the information on the photographic plate). In other words, he is suggesting that the brain does not store information like a digital computer but deals with interference patterns in three dimensions. So, on this theory, the brain receives signals from the universe consisting of energy at many different frequencies. It transforms these signals to make objects out of blurs or frequencies, making them into sounds and colours, movements, smells and tastes. In reality then, according to Pribram, there is no space and no time, just events. The brain projects the world 'out there', acting rather like a stereoscope.

Combining the ideas of these two thinkers, it could be said that the brain, according to their views, is the basic cause of the physical world being as it is. The brain is abstracting from the undivided wholeness; it is unfolding the world's phenomena from the enfolded unified universe. So the brain is a 'frequency analyser'.

So all our world, according to Bohm — including us and our experiences, thoughts and emotions — are enfolded within the overall order or reality. Time and space are enfolded in it, too, mixed with everything else. This is the undivided reality, the infinite plenum, of the mystics. The brain abstracts from it the physical world, which is therefore a *maya*, a great illusion. Transcendental experiences — mystical states — may allow us occasional means of access to that realm of undivided totality. In other words, by-passing our normal, constricting perceptual mode (what Aldous Huxley calls the reducing valve of the brain) may attune us to the source, or matrix, of reality. Very briefly then, Bohm's holographic theory says that our brains mathematically construct hard reality by interpreting frequencies transcending time and space. 'The brain is a hologram, interpreting a holographic universe.'

Pribram reassures us by saying that he does not 'understand' any of this. It is not possible to apply normal linear logical thought processes to a region to which they do not apply. It is not possible to understand all this. One's very thoughts about it are abstractions from it. The only way to understand is to become one with it – as we are all the time in reality. Arguing about the mass and velocity of that ink spot in Bohm's analogy is not likely to lead to a deep and consistent knowledge because that is not its basic nature. Electrons are sometimes waves, sometimes particles, jumping from point to point, because they are in reality neither wave nor particle. These two forms are abstracts, with the rest of the experiment, and the observer, and his or her thought about it. All form a unified whole with the rest of the universe. According to Bohm (if I understand him correctly) quantum mechanics is a mathematical description of the implicit universe.

Based on this idea then, individual brains are bits of the greater hologram. They have access under certain conditions to all the information in the total system. So we have in principle an explanation (if you can call it that) for all the psychic phenomena. Regarding telepathy, the thought is everywhere and does not have to be transmitted. On this theory no energy transmitting it will ever be found. Psychokinesis, the mind's effect on matter, may be a natural result of interaction at the primary level. Healing, clairvoyance, all fit in, occurring in a dimension transcending time and space.

Sheldrake

Now what of Sheldrake? Rupert Sheldrake, a biologist, has produced a theory which is tentatively suggesting why living forms have the same shape and characteristics as others in the same species. Most biologists appear to think that bodily characteristics – and also instincts and tendencies (for example, the web-building 'knowledge' of spiders) – are all in the genetic code but that learned behaviour cannot be inherited. Certain experimental facts appear to contradict this and indicate a unifying link with others and with the past. The English pioneer of experimental psychology, William McDougall, did experiments in the United States in which rats learned to escape from a tank by choosing a dark exit rather than a light exit: the latter, when they chose it, gave them an electric shock. He counted the number of trials before the rats learned to take the correct exit. This experiment was repeated in the United Kingdom and it was found that the rats learned how to escape more quickly. It was repeated once more in Australia and there they learned to escape even more quickly. Sheldrake suggests that this is surely evidence that they are all linked together in a non-physical way.

There is clearly a great deal more to the formation and behaviour of animals than is to be found in the genetic code, that is, in the genes in the chromosomes. The genetic programme cannot be the same as the chemical structure of DNA because all cells of the body can turn out identical copies of DNA yet they develop differently. Some influential biologists think – especially Sheldrake – that the developing limbs and organs of embryos are shaped by 'morphogenetic fields' (form-generating fields) which connect similar things together across space and across time. The embryo, as it were, tunes in to the form of past members of the species. In their instinctive behaviour animals draw on a sort of memory bank or pooled memory of their species.

If this is in principle true – and it has been described, or something rather like it, for a very long time indeed in the East – then it could explain in principle why all human beings have, by and large, somewhat similar experiences of the physical world. (I should add that Sheldrake's modern version of an ancient Hindu theory has received considerable technical criticism. However, it is a good theory in that it is testable – and some tests have already proved positive.) It may be that the physical world as apparently perceived by human beings has been gradually built up until it has assumed its present form as a result of the successive gradual changes in the structure of the brain. If the brains of present human beings are the results of 'subtle influences' from past and present human beings during the brain's embryonic development, then this would explain why all human beings appear to see the same major features of the physical world. (I have already mentioned that a high proportion of the things we see around us have been put there by us, in 'making sense' of our perceptions, as any psychologist would confirm.)

One is more or less pushed into the idea that there must be some forming force or intelligence behind the system doing all this.

CONCLUSIONS

So what can, in summary, be said of these ideas? First, the holographic theory, as does any good hypothesis, raises urgent and new questions. Implicit in it is the assumption that harmonious coherent states of consciousness are more nearly attuned to the primary level of reality. The question that naturally arises is, What is it that is fragmenting us? Even our normal language, in which we describe everything, is based upon fragmentation. We need a new language based on unity, on a universal unified field of being. Bohm has a chapter in his book on language and the difficulties that arise as a result of our present language. The knower and the known are constructs; in the implicate world all is one and there is no separation of space or time, knower and known. The whole point of the manifest world we recognize is, it appears to have separate units; that is why we have the kind of brains that we have, separate but interacting. But deep down the consciousness of mankind is one. We construct space and time for our own convenience — or rather our inherited brains appear to do so. Somehow we have to try to leave thought behind, transcending it, and meditation is a way of loosening the hold it has upon us. Meditation, Bohm says, transforms the mind, transforms consciousness and leads to what he calls insights. This in turn may even lead to a change in the physical structure of the brain. As mentioned earlier (page 89) the brain structure of a psychic does seem to be slightly different from that of the ordinary non-psychic person.

The consciousness researcher, John Lilly, has relevant things to say about the structure of the brain — or rather, as he puts it, about the 'programming of the brain'. Lilly did experiments on himself in sensory deprivation and, originally, also under the effects of LSD. Lilly was able to 'leave the body' and 'go into various spaces', as he describes it. He speaks of being 'programmed' for various spaces. Lilly could then go to any kind of a world he could imagine. One of his rules is, 'What one believes to be true either is true or becomes true in one's mind, within limits to be determined experimentally or experientially.... The limits are beliefs to be transcended.' Lilly speaks of the physical world as 'ordinary concensus reality'. Lilly was able 'to find Great Beings, points of radiance, teaching him, advising him' telling him to go back to his body 'eventually to perceive the oneness of them, of him, and of others'. They said they were his guardians. He looked upon them as 'two aspects of his own functioning at the supra-self level' (like Jung's Philemon or Socrates' daemon). Alternatively they may be, he says, 'entities in other spaces, other universes, than our concensus reality'. They may be 'helpful constructs, helpful concepts' he uses for his own evolution. They may be representatives of an esoteric school. They may be concepts functioning in his own human biocomputer at the supraspecies level. They may be members of a civilisation one-hundred thousand years or so ahead of ours. They may be a tuning in on two networks of communication of a civilisation way beyond ours, radiating information throughout the galaxy.

Lilly suggests you choose whatever seems right to you! How is it possible to know? I propose to leave his very stimulating views at that, reminding the reader that at a very high level of consciousness all life appears to be one. A final thought: Lilly found, as he puts it, 'a sort of zero point of blackness and silence from which he could move out in any direction he could conceive.' His claim is that: *anything that could be imagined, exists.*

Pribram says, 'The order in the universe appears to be so indistinguishable from the mental operations by which we operate on that universe that we must conclude either that our science is a huge mirage, a construct of our convoluted brains, or that, indeed, as proclaimed by all our great religious convictions, a unity characterizes this emergent, and is the basic order of the universe.' Bohm also says that the idea of the knower observing the known across the gulf of unknowing must be replaced by the paradigm of a unified field of being, a self-conscious universe realizing itself to be integrally whole and interconnected. Knower and known are thus, on this view, falsehoods, crude constructs based on abstraction. In the non-manifest implicate order, all is one. There is no separation of space and time. The whole point of the manifest world is, it seems to me, to have separate units — separate but interacting. In nonmanifest reality it is all interpenetrating, interconnected in one. *Deep down the consciousness of mankind is one. We construct space and time for our own convenience.*

So what is the value of this idea of an implicate/explicate universe, besides its relevance to science. Bohm says that it is a sort of bridge. Meditation can bring us along it. If we consider it seriously, apart from its utility in understanding matter (science), the bridge (or pier) would help us to loosen our way of considering consciousness so that it does not hold us so rigidly. Somewhere, somehow, 'we have got to leave thought behind and come to this emptiness of this manifest thought and the conditioning of the non-manifest mind by the seeds of manifest thought'. So, Bohm says, meditation actually transforms the mind: it transforms consciousness. It leads to 'insight' — a physical change in the brain structure — a change in the 'programming' which hitherto had led to this physical world being considered as the only reality.

Finally, the Bohm holomovement is the source of all possibilities. What is taken out of it depends on us, and on what we do. Holding clear thought images tends to actualize them. But we are also integral with the holographic universe. So our thoughts and emotions, our brains and their programmes, are all abstracted. Every part of everything is in every other part.

How impossible it is to understand all this with the rational mind — but the evidence does appear to point in this direction.

A final thought or two.... My own experience so far does indicate that the result of quiet meditation, without thought, is a coherence, a state of order, all over the brain, indicated by the brain rhythms — the EEG activity. This has been claimed as one result of transcendental meditation, but it seems to me certainly not to be unique to that type of meditation. Quietly listening to music may lead to the same orderly healing state in which all the bodily systems tend to revert to normal.

I think that perhaps Bohm and Pribram have started a major paradigm shift, applicable to the whole of science, life, and everything else. I wonder if you agree! We live in exciting times!

Religious teachers, from firsthand experience or an insight into this greater reality, tell us to behave in certain ways — unselfish ways. The great saints — of all religions — are utterly unselfish, thinking only of the good of others and having no desires of their own. Our way of development is perhaps in this direction, away from the separate self and separate objects to the unity from which we perhaps came. The great emphasis of the religions, especially Christianity, on love, as the most fundamental 'force' of the universe would then be clear, as the supreme expression (at least in its finest form) of the loss of self and the overall importance of unity. The battle between good and evil is perhaps to be looked upon as love and unity versus selfishness and separateness, evolution versus devolution.

So if there is one 'message' suggested to us by the evidence in the preceding pages it is perhaps one of the most ancient: Know thyself! Seek within! All the mysteries and anomalies we have considered appear to have their solutions in the mind. Through meditation we can first train the mind and then perhaps transcend it — and know! But that knowledge will not be expressible in words.

The world appears to be an 'educational toy', perhaps designed to bring us to this position — traditionally, from the Unreal to the Real. That at least is how it appears to me. Now it is up to you. Perhaps through this book, perhaps through other sources, including the great scriptures of the world, you may discover a new understanding, a new world. There are so many exciting discoveries to be made, both in science and the 'physical world', and within ourselves — especially within ourselves — and in the unity, the whole, from which all duality springs.

BIBLIOGRAPHY and GLOSSARY

FURTHER READING

BLACKMORE, S.J. *Beyond the Body,* UK: Heinemann, 1982.

BOHM, D.J. *Wholeness and the Implicate Order,* UK: Routledge and Kegan Paul, 1980.

GREY, M. *Return from Death,* UK: RKP Arcana, 1985.

HASTED, J. *The Metal Benders,* UK: Routledge and Kegan Paul, 1981.

HAYNES, R. *The Hidden Springs,* UK: Hollis and Carter, 1961.

HEYWOOD, R. *Beyond the Reach of Sense,* USA: Dutton, 1974.

INGLIS, B. *Natural and Supernatural,* UK: Hodder and Stoughton, 1977.

INGLIS, B. *Science and Parascience,* UK: Hodder and Stoughton, 1984.

JAHN, R.G. and DUNNE, B.J. *Margins of Reality* USA: 1987

LeSHAN, L. *The Medium, the Mystic, and the Physicist,* UK: Turnstone, 1974.

LeSHAN, L. *The Science of the Paranormal,* UK:Aquarian Press, 1987

MACKENZIE, A. *Hauntings and Apparitions,* UK: Heinemann, 1982.

MANNING, M. *The Link,* UK: Colin Smythe, 1974.

MONROE, R.A. *Journeys Out of the Body,* USA: Doubleday, 1971.

MOODY, M. *Life After Life,* UK and USA: Bantam, 1976.

OWEN, I.M. and SPARROW, M. *Conjuring up Philip,* Canada: Fitzhenry and Whiteside, 1976.

SABOM, M.B. *Recollections of Death,* UK: Corgi, 1982.

SHELDRAKE, R. *A New Science of Life,* UK, USA: Granada, 1981.

TARG, R. and HARARY, K. *The Mind Race,* UK: New English Library, 1985. USA: Villard Books, 1984.

TARG, R. and PUTHOFF, H.E. *Mind-Reach,* USA: Delacorte Press/Eleanor Friede, 1977.

WHITEMAN, J.H.M. *The Meaning of Life,* UK: Colin Smythe, 1986.

MAJOR SOURCES

BESANT, A. and LEADBEATER, C.W. *Occult Chemistry,* UK: Theosophical Publishing House, 1919.

BESANT, A. and LEADBEATER, C.W. *Science of the Sacraments,* India: Theosophical Publishing House, 4th edition 1957.

BERSTEIN, M. *The Search for Bridey Murphy,* N.Y.: Doubleday 1956.

COURTIER, J. *Rapport sur les séances d'Eusapia Palladino à l'Institut Général Psychologique en 1905, 1906, 1907 et 1908,* France: Institut Général Psychologique, 1908.

CRAWFORD, W.J. *The Reality of Psychic Phenomena,* UK: Watkins, 1916.

CRAWFORD, W.J. *Experiments in Psychical Science,* UK: Watkins, 1919.

CUMMINS, G. and CONNOLL, R. *Perceptive Healing,* UK: Rider, 1945.

DAVEY, S.J. *The possibilities of mal-observation and lapse of memory from a practical point of view,* UK: Proc. SPR Vol. 4, 1887.

DUNNE, J.W. *An Experiment with Time,* USA: Macmillan, 1927.

FENWICK, P. et al. *Lucid Dreaming: correspondence between dreamed and actual events in one subject during REM sleep,* Netherlands: Biological Psychology, Vol. 18, 1984.

FENWICK, P. et al. *'Psychic sensitivity', mystical experience and brain pathology,* UK: British Journal of Medical Psychology, Vol. 58, 1985.

GARRETT, E.J. *My Life as a Search for the Meaning Of Mediumship,* UK: Rider, 1939.

GLASKIN, G.M. *Windows of the Mind: The Christos Experience,* UK: Wildwood House, 1974.

GREEN, G.E. *Lucid Dreams* UK: Inst. of Psychophysical Research, 1968.

GURNEY, E., MYERS, F.W.H. and PODMORE, R. *Phantasms of the Living,* UK: Trübner, 1886. Two vols.

IVERSON, J. *More Lives than One?* UK: Souvenir Press, 1976.

JAHN, R.G. *The Persistent Paradox of Psychic Phenomena: an Engineering Perspective,* USA: Proc. IEEE, Vol. 70, 1982.

LEANING, F.E. *An Introductory Study of Hypnagogic Phenomena,* UK: Proc. SPR, Vol. 35, 1925.

MULDOON, S. & CARRINGTON, H. *The Projection of the Astral Body.* UK: Rider, 1929.

MYERS, F.W.H. *Human Personality and Its Survival of Bodily Death.* UK: Longmans Green, 1903, 2 vols.

NORTHROP, F.S.C. *The Logic of the Sciences and the Humanities,* USA: MacMillan, 1947.

OWEN, A.R.G. *Can We Explain the Poltergiest?* USA: Garret Publications, 1964.

PATANJALI, *How to Know God: The Yoga Aphorisms of Patanjali. Translation and commentary by Prabhavananda & Isherwood, C.* UK: George Allen and Unwin, 1953.

PRICE, H.H. *Survival and the Idea of 'Another World',* UK: Proc. SPR, Vol. 50, 1953-56.

SALTMARSH, H.F. *Evidence of Personal Survival from Cross Correspondences,* UK: Bell, 1938.

SCHMEIDLER, G.R.'*In Handbook of Parapsychology, Ed. Wolman, B.B.* USA: Van Nostrand Reinhold, 1977.

SOAL, S.G. *A Report of some Communications Received through Mrs Blanche Cooper,* UK: Proc. SPR Vol. 35, 1925.

SPINELLI, E. *Human Development and Paranormal Cognition,* UK: Ph.D. thesis, University of Surrey, 1978.

STEVENSON, I. *Twenty Cases Suggestive of Reincarnation,* US: University Press of Virginia, 1974.

TYRRELL, S.N.M. *Apparitions,* UK: MacMillan, 1962

WHITEMAN, J.H.M. *The Mystical Life,* UK: Faber and Faber, 1961.

WOLMAN, B.B. (Ed.) *Handbook of Parapsychology,* USA: Van Nostrand Reinhold, 1977.

YRAM, *Practical Astral Projection,* USA: Samuel Weiser, 1972.

GLOSSARY

Apport Object which appears paranormally during a séance
Clairaudience The 'hearing' of hallucinatory sounds and voices
Clairvoyance 'Seeing' hallucinatory visions. Information gained directly from an inaccessible object, not from someone's mind
Cryptomnesia A memory consciously forgotten but manifested in conscious experience
Dissociation Altered state of consciousness in which the subject is not fully aware or analytical
Distant viewing/remote perception Receiving mental impressions that appear to relate to surroundings experienced by a distant agent
Dualism Philosophical view that reality consists of two principles (usually, mind and matter)
Ectoplasm Substance emanating from bodily orifices of a physical medium — may 'materialize'
EEG (Electroencephalograph) Machine that amplifies and records the voltages between electrodes fixed to the scalp
EMG(Electromyograph) Machine that amplifies and records voltages between electrodes fixed to the skin over muscles
ESP (Extrasensory perception) Aquisition of information other than through the five senses
George Author's term for the personal unconscious mind that retrieves memories, dramatizes dreams, makes ESP information overt and responds to hypnosis
Gestalt A wholeness: all the various elements of a situation being necessary to produce given result
Guide Spiritualist term: a discarnate entity who helps transmit information from other levels of consciousness. A 'control' may take charge of medium's body
Hallucination Apparent sensory perception with no physical cause
Hypnagogic While falling asleep
Hypnopompic While waking up
Materialization Tangible form ostensibly made from ectoplasm
Naive realism Philosophical conception that the objects and space of the physical world and the progression of time are independent of the observer, that they may be examined without being altered
OOBE (Out-of-the-body-experience) Consciousness appears to be in a point in space away from the physical body
Paradigm A mental model or pattern that helps to make sense and order of experience
PrecognitionESP of future events
Post- or retrocognition ESP of past events
Psychokinesis (PK) Paranormal movement of an object
Reductionist medicine When patient is regarded predominantly as an electro-chemical machine

INDEX